Plasmids

The Practical Approach Series

D. RICKWOOD
Department of Biology, University of Essex
Wivenhoe Park, Colchester, Essex CO4 3SQ, UK

B. D. HAMES
Department of Biochemistry and Molecular Biology
University of Leeds, Leeds LS2 9JT, UK

Affinity Chromatography

Anaerobic Microbiology

Animal Cell Culture
(2nd Edition)

Animal Virus Pathogenesis

Antibodies I and II

Behavioural Neuroscience

Biochemical Toxicology

Biological Data Analysis

Biological Membranes

Biomechanics—Materials

Biomechanics—Structures and
Systems

Biosensors

Carbohydrate Analysis

Cell–Cell Interactions

The Cell Cycle

Cell Growth and Division

Cellular Calcium

Cellular Interactions in
Development

Cellular Neurobiology

Centrifugation (2nd Edition)

Clinical Immunology

Computers in Microbiology

Crystallization of Nucleic Acids
and Proteins

Cytokines

The Cytoskeleton

Diagnostic Molecular Pathology
I and II

Directed Mutagenesis

DNA Cloning I, II, and III

Drosophila

Electron Microscopy in Biology

Electron Microscopy in
Molecular Biology

Electrophysiology

Enzyme Assays

Essential Developmental
Biology

Essential Molecular Biology I
and II

Experimental Neuroanatomy

Fermentation

Flow Cytometry

Gas Chromatography

Gel Electrophoresis of Nucleic
Acids (2nd Edition)

Plasmids

A Practical Approach

Edited by

K. G. HARDY

Glaxo Institute for Molecular Biology
14 Chemin des Aulx
1228 Plan-les-Ouates
Geneva, Switzerland

OXFORD UNIVERSITY PRESS
Oxford New York Tokyo

Oxford University Press, Walton Street, Oxford OX2 6DP

Oxford   New York   Toronto
Delhi   Bombay   Calcutta   Madras   Karachi
Kuala Lumpur   Singapore   Hong Kong   Tokyo
Nairobi   Dar es Salaam   Cape Town
Melbourne   Auckland   Madrid
and associated companies in
Berlin   Ibadan

Oxford is a trade mark of Oxford University Press

A Practical Approach 🛈 is a registered trade mark
of the Chancellor, Masters, and Scholars of the University of Oxford
trading as Oxford University Press

Published in the United States
by Oxford University Press Inc., New York

A catalogue record for this book is available from the British Library

Library of Congress Cataloging in Publication Data
Plasmids : a practical approach / edited by K.G. Hardy.–2nd ed.
(The Practical approach series)
Includes bibliographies and index.
1. Plasmids–Research–Methodology.   I. Hardy, K.G. (Kimber G.)
II. Series.
QAR76.6.P56   1993   589.9'08734–dc20   93–14282
ISBN 0–19–963445–9 (h/b)
0–19–963444–0 (p/b)

Typeset by Footnote Graphics, Warminster, Wilts
Printed in Great Britain by Information Press Ltd, Eynsham, Oxon

Preface

This book is designed to provide research workers with complete and detailed protocols for studying bacterial plasmids and for using both plasmids and phagemids as vectors.

The techniques for studying plasmid replication and maintenance in both Gram-negative and Gram-positive bacteria are described in detail, as are fundamental methods for purifying plasmids, and for introducing them into cells. Also included are protocols for mutagenesis and for analysing plasmid-encoded products.

Other chapters focus on techniques for studying plasmids and for using plasmid vectors in particularly important groups of bacteria: animal and plant pathogens harbouring virulence plasmids, *Streptomyces* and lactococci.

λ−plasmid composite vectors and phagemids of various kinds are extremely useful for expression cDNA cloning and other genetic manipulations. The protocols for using them, as well as important yeast vectors, are described in detail.

I am delighted that so many leading researchers have agreed to contribute a chapter and I am sure their experimental protocols will stimulate research on plasmids as well as the use and development of plasmid and phagemid vectors.

Geneva   K.G.H.
July 1993

Contents

2. Plasmids from Gram-positive bacteria 39

Juan C. Alonso and Manuel Espinosa

3. Lactococcal plasmid vectors 65

K. J. Leenhouts and G. Venema

6. Streptomyces plasmid vectors

W. Wohlleben and G. Muth

7. Use of λ–plasmid composite vectors for expression cDNA cloning

Toru Miki and Stuart A. Aaronson

8. Phagemids and other hybrid vectors   197

Michelle A. Alting-Mees, Peter Vaillancourt, and Jay M. Short

Contents

Contributors

LUIS A. ACTIS
Department of Microbiology and Immunology, Oregon Health Sciences University, L220 3181, S.W. Sam Jackson Park Road, Portland, Oregon 97201, USA.

STUART A. AARONSON
Bldg 37-1E24, Laboratory of Cellular and Molecular Biology, National Cancer Institute, Bethesda, MD 20892, USA.

JUAN C. ALONSO
Max-Planck-Institut für molekulare Genetik, Ihnestrasse 73, D 1000 Berlin 33, Germany; and Centro Nacional de Biotecnología, CSIC, Campus de Cantoblanco, 28049 Madrid, Spain.

MICHELLE A. ALTING-MEES
STRATAGENE, 11099 North Torrey Pines Road, La Jolla, CA 92037, USA.

JORGE H. CROSA
Department of Microbiology and Immunology, Oregon Health Sciences University, L220 3181, S.W. Sam Jackson Park Road, Portland, Oregon 97201, USA.

ROLF DEBLAERE
Laboratorium voor Genetica, Universiteit Gent, K.L. Ledeganckstraat 35, B-9000 Gent, Belgium.

JAN DESOMER
Laboratorium voor Genetica, Universiteit Gent, K.L. Ledeganckstraat, 35, B-9000 Gent, Belgium.

MANUEL ESPINOSA
Centro de Investigaciones Biológicas, CSIC, Velazquez 144, 28006 Madrid, Spain.

K. J. LEENHOUTS
Department of Genetics, Center for Biological Sciences, University of Groningen, Kerklaan 30, 9751 NN, The Netherlands.

TORU MIKI
Bldg 37-1E24, Laboratory of Cellular and Molecular Biology, National Cancer Institute, Bethesda, MD 20892, USA.

G. MUTH
Abteilung für Angewandte Molekularbiologie, Fachbereich Biologie, Universität des Sarrlandes, Bau 2, D-66041 Saarbrücken, Germany.

Contributors

KURT NORDSTRÖM
Department of Microbiology, Uppsala University, Biomedical Center, 751 23 Uppsala, Sweden.

JAY M. SHORT
STRATAGENE, 11099 North Torrey Pines Road, La Jolla, CA 92037, USA.

ROBERT S. SIKORSKI
Massachusetts General Hospital, Department of Medicine, Harvard University School of Medicine, Fruit Street, Boston, MA 02114, USA.

MARCELO E. TOLMASKY
Department of Microbiology and Immunology, Oregon Health Sciences University, L220 3181, S.W. Sam Jackson Park Road, Portland, Oregon 97201, USA.

PETER VAILLANCOURT
STRATAGENE, 11099 North Torrey Pines Road, La Jolla, CA 92037, USA.

MARC VAN MONTAGU
Laboratorium voor Genetica, Universiteit Gent, K.L. Ledeganckstraat 35, B-9000 Gent, Belgium.

G. VENEMA
Department of Genetics, Center for Biological Sciences, University of Groningen, Kerklaan 30, 9751 NN, The Netherlands.

W. WOHLLEBEN
Abteilung für Angewandte Molekularbiologie, Fachbereich Biologie, Universität des Sarrlandes, Bau 2, D-66041 Saarbrücken, Germany.

Abbreviations

ADC	automatic directional cloning
Ap	ampicillin
ARS	autonomously replicating sequence
Bc	bacteriocin production
BHR	broad host range
CAT	chloramphenical acetyltransferase
CCC	covalently closed circular
cDNA	complementary DNA
CEN	centromere sequence
Cm	chloramphenicol
DEPC	diethyl pyrocarbonate
DMSO	dimethylsulfoxide
ds	double-stranded
DTT	dithiothreitol
EDTA	ethylenediamine tetraacetic acid
Em	erythromycin
EtBr	ethidium bromide
FGF	fibroblast growth factor
Fi plasmid	fasciation-inducing plasmid
Gm	gentamycin
Hy	haemolysin
IPTG	isopropylthio-β-D-galactoside
KGF	keratinocyte growth factor
Km	kanamycin
LB	Luria–Bertani
md	megadaltons
MCS	multiple cloning site
MES	multiple excision site
MIC	minimal inhibitory concentration
MOPS	3-(N-morpholino) propanesulfonic acid
Nal	nalidixic acid
NHR	narrow host range
Nm	neomycin
ONPG	O-nitrophenyl-β-D-galactoside
PAGE	polyacrylamide gel electrophoresis
PEG	polyethylene glycol
p.f.u.	plaque-forming units
RC	rolling circle
Ri	root-inducing plasmid

RNasin	RNase inhibitor protein
Sarkosyl	sodium sarcosinate
SDS	sodium dodecyl sulfate
ss	single-stranded
SSC	standard saline citrate
Ti plasmid	tumour-inducing plasmid
Tc	tetracycline
TCA	trichloroacetic acid
ts	temperature-sensitive
VBHR	very broad host range
X-gal	5-bromo-4-chloroindolyl-β-D-galactoside

1

Plasmid replication and maintenance

KURT NORDSTRÖM

1. Introduction

All the essential genetic information in bacteria is located in one double-stranded, circular DNA molecule, the chromosome. In addition, it is the rule, rather than the exception, that bacteria contain other DNA molecules, plasmids, that live in harmony with their hosts. These are separate entities, replicate autonomously, and control their own replication. They are present in defined copy numbers (average number of molecules per cell) during exponential growth of their hosts. The copy number is determined by genes present on the plasmids but is also affected by the host and by growth conditions (1). In this chapter, replication of plasmids as well as the control of replication will be discussed, with particular emphasis on plasmids that replicate in *Escherichia coli*.

The smallest part of a plasmid that is able to replicate with the same copy number as that of the full-size plasmid is called the basic replicon. It is normally about 2 kb and contains one (in some cases two or three) origins (*ori*) of replication, codes for one (in some cases two or three) proteins that are involved in the initiation of replication, and carries genetic information for the control of replication (1) (*Figure 1* and *Table 1*).

2. Single-cell resistance (SCR) to antibiotics

Many plasmids carry genes for antibiotic resistance, which is a quantitative phenotype (2). The resistance level can be measured in different ways. One is to incubate fairly large inocula in the presence of increasing amounts of the drug. The lowest concentration that totally inhibits growth is called the MIC (minimum inhibitory concentration). However, in genetic work, single-cell resistance (SCR) is a more useful measure. A few hundred bacteria are spread on plates containing different concentrations of the drug. The number of colonies is counted after incubation overnight (*Figure 2*). Up to a certain level (SCR) all cells plated give rise to colonies. Then the number of colonies

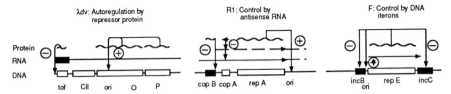

Figure 1. Basic replicons. One example of each class of plasmid replicons in which replication is regulated by a repressor protein (λdv), an antisense RNA (R1), and DNA iterons (F), respectively. The functions of the various substances is denoted, − and + for inhibition and requirement, respectively. The F iterons are located in two clusters, *incB* and *incC*, respectively. The RepE protein of plasmid F is used positively in initiation of replication but inhibits its own synthesis.

Table 1. Basic replicons

Plasmid	Size of basic replicon (kb)	Number of origins	Cop function	Rep protein
ColE1	1	1	RNAI Rop protein	None
R1 (IncFII)	2	1	CopA RNA CopB protein	RepA
pT181	1.5	1	RNA	RepC
λdv	2	1	Cro protein	O, P
P1	2	1	Iterons	RepA
F	2	2	Iterons	RepE
R6K	4	3	Iterons	π
RK2	3	1	Iterons	TrfA

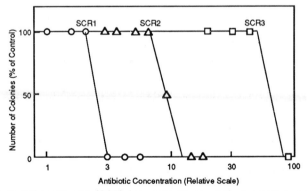

Figure 2. Single-cell resistance (SCR) (see also *Protocol 1*). A small number of exponentially growing cells are spread on plates containing different amounts of an antibiotic. The number of colonies are counted after incubation overnight. The highest concentration at which all cells plated give rise to colonies is indicated and denoted the SCR. The graph shows a hypothetical result with three different plasmids.

drops and at an antibiotic concentration about 50% higher than SCR no colonies are formed. At antibiotic concentrations just above SCR, the colony size becomes smaller and variable. The method (*Protocol 1*) is highly reproducible and a 15% difference in SCR values between two strains is highly significant and reproducible.

Protocol 1. Determination of single-cell resistance (SCR) to antibiotics

1. Grow the bacteria exponentially in LB or minimal medium.

2. At a cell density of about 10^8/ml, dilute the population 10^5-fold in 0.9% (w/v) NaCl.

3. Spread 0.1 ml of the dilution on plates containing the same medium as the growth medium and different concentrations of the antibiotic to be tested. The concentrations should differ by $\approx$1.5-fold or less, (e.g. 0, 50, 70, 100, 150, 200, 300, 400, 600, 800, etc., µg/ml).

4. Incubate the plates overnight at 37°C or other appropriate temperature.

5. Count the number of colonies and note whether the colony size is uniform or varied.

6. Plot the number of colonies as a function of the concentration of the antibiotic (the latter should be a logarithmic scale). The highest concentration at which all cells plated give rise to colonies is the single-cell resistance (SCR) level (see *Figure 2*).

SCR is a quantitative phenotype that has proven useful in studies of many different systems. The fact that bacteria have defined resistance levels is extremely valuable in the selection of mutants (see below).

Resistance is dominant and generally due to enzymes that metabolize the drugs. Expression of the genes is often constitutive and, hence, in these cases specific activity of the enzymes is proportional to the gene dosage. Therefore, the resistance level increases with increasing gene dosage (*Figure 3*). For some antibiotics (e.g. penicillins), there is a linear correlation between resistance and gene dosage over a very large range, whereas for other antibiotics, resistance increases with gene dosage until it reaches a plateau (2). For chloramphenicol, this is due to the very large energy consumption required for acetylation of the drug, whereas the specific activity of the enzymes that metabolize streptomycin and kanamycin is very low but the drugs seem to reduce the efficiency of the outer penetration barrier; increasing amounts of the enzyme, therefore, can not compensate for the increased rate of uptake of the antibiotics (2, 3).

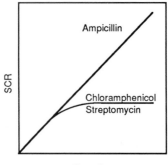

Figure 3. Antibiotic resistance as a function of the dosage of the enzyme that metabolizes the antibiotic.

3. Determination of plasmid copy numbers

3.1 Copy numbers

3.1.1 Absolute and relative copy numbers

Copy numbers are given either as numbers of molecules per chromosome equivalent or per cell. The choice of measure depends on the purpose of the determination. Copy numbers per chromosome equivalent are easier to determine than copy numbers per cell. Easiest to determine are relative copy numbers, i.e. copy numbers of a plasmid compared to that of a reference plasmid (e.g. the wild-type).

Copy number per cell can be calculated directly by measuring the amount of plasmid DNA in a population, the size of which has been determined by counting the bacteria microscopically in a counting chamber or electronically in a Coulter counter or flow cytometer.

Relative copy numbers are given either as numbers of plasmid copies per genome equivalent (4.8 Mb in *Escherichia coli*) or normalized to the copy number of the wild-type plasmid.

3.1.2 Chromosomal DNA per cell

Determination of the number of copies per cell from ratios between plasmid and chromosomal DNA requires that the amount of chromosomal DNA per cell is known. The latter can be determined directly by measuring the amount of DNA in a known population of bacteria. The size of the population can be determined by using counting chambers, Coulter counters, or flow cytometry. Alternatively, the amount of chromosomal DNA can be estimated in the following way. Cooper and Helmstetter (4) deduced that the average number of chromosome equivalents (G) per cell in an exponentially growing population of bacteria is given by:

$$G = (\tau/C \cdot \ln 2)[2^{(C + D)\tau} - 2^{D/\tau}]$$

where

> τ = generation time
> C = the time required for one complete round of replication
> D = the time between the end of one round of replication and cell division.

Hence, if C, D, and τ are known, G can be calculated. This value and the ratio between plasmid and chromosomal DNA give the number of plasmid copies per cell (5).

The number of plasmid copies per cell can also be estimated by comparing the gene dosage of a gene on the plasmid and the same gene present on the chromosome, e.g. the β-lactamase gene of Tn3 inserted as λ:Tn3 in the normal λ attachment site on the chromosome and Tn3 transposed on to the plasmid (5). Collins and Pritchard (6) deduced that the average number of copies (F) per cell of a particular gene in an exponentially growing population of bacteria is:

$$F = 2^{[C(1-x)+D]/\tau}$$

where x is the relative distance between the origin of replication and the gene ($x = 1$ defines the terminus of replication).

It should be pointed out that SCR is particularly suitable for determining copy numbers of plasmids that are not stably maintained because they lack partition (*par*) and other maintenance functions (see Section 4.1).

3.2 Amount of plasmid DNA

The amount of plasmid DNA can be measured in two fundamentally different ways; by direct determination and by determining the dosage of an antibiotic resistance gene carried on the plasmid (see above). Other constitutive genes can also be used for determining gene dosages.

3.2.1 Direct determination of plasmid DNA

The amount of plasmid DNA can be determined by density gradient centri- fugation (7) (*Protocol 2*) or by gel electrophoresis (by standard procedures (8)). Both methods require an internal standard which is often the amount of chromosomal DNA. With radioactively labelled DNA, the relative amount of plasmid DNA can be determined by scanning of the gels, e.g. in a Phosphor- Imager (Molecular Dynamics, Sunnyvale, CA).

Protocol 2. Quantitative determination of plasmid DNA and separation of plasmid and chromosomal DNA by CsCl/ethidium bromide centrifugation

A. *Quantitative assay*

1. Grow the bacteria exponentially in 10 ml of LB or minimal medium (containing 250 μg/ml of adenosine) to a cell density of about 5×10^7/ml.

Protocol 2. *Continued*

2. Add 50 μl of [³H]thymidine (20 Ci/mmol, 1 mCi/ml). Continue growth to a cell density of about 2×10^8/ml.

3. Chill the culture on ice. Harvest the cells by centrifugation. Wash the cells twice with cold TES (20 mM Tris–HCl, 1 mM EDTA, 20 mM NaCl, pH 7.5) buffer. Resuspend in 2 ml cold TES buffer.

4. Add lysozyme (to a final concentration of 1 mg/ml) and RNase (to a final concentration of 500 μg/ml). Incubate for 15 min at 37°C.

5. Add Sarcosyl to a final concentration of 1%. Shear the lysates by sucking five times through a 21 gauge syringe; after this, the lysate should be non-viscous.

6. Mix 2 ml of lysate with 3 ml of TES buffer. Add 5 g of CsCl. Add 0.5 ml of ethidium bromide solution (5 mg/ml). The refractive index should be 1.4055. This CsCl concentration is correct for *E. coli* DNA—for bacteria with other GC contents, the appropriate CsCl concentration is different.

7. Centrifuge at 40 000 r.p.m. in a Ti75 or Ti50 rotor in a Beckman 75 ultracentrifuge for 40 h. In a vertical rotor, a much shorter centrifugation time (2–6 h) is sufficient.

8. Collect the gradients in about 40 fractions. Precipitate aliquots from each fraction in 10% TCA (trichloroacetic acid). Count the radioactivity in each sample in a liquid scintillation counter.

B. *Separation of plasmid DNA from chromosomal DNA*

In this case, no radioactive labelling is necessary.

1. Grow the bacteria exponentially to a cell density of about 10^8/ml.

2. Harvest and treat the bacteria either as described in steps **3–7** above or prepare so-called 'cleared lysates' as in C below that are then centrifuged as described in steps **6–7** above.

3. The plasmid and the chromosomal DNA bands can be visualized by UV illumination. Collect the plasmid band.

C. *Preparation of cleared lysates (58)*

1. Harvest the bacteria by centrifugation.

2. Resuspend the pellet in 25% of sucrose in Tris–HCl (0.25 M, pH 8.0) to give a 50-fold increase in the concentration of the cells compared to the culture.

3. Mix well by stirring. Add one volume of a solution of lysozyme (10 mg/ml) containing RNase (1 mg/ml) to five volumes of sample, (e.g. 10 μl to 50 μl).

4. Place on ice for 5 min. Add two volumes of EDTA (0.25 M, pH 8.0).

5. Incubate on ice for 5 min. Add eight volumes of lysis buffer (50 mM Tris–HCl, 63 mM EDTA, 1% Brij-58, 0.4% deoxycholate, pH 8.0).

6. Incubate on ice for about 5 min or until the samples become viscous.

7. Spin down the debris by centrifugation (20 000 *g* for 20 min). Save the supernatant for analysis.

3.2.2 Determining copy number by measuring gene dosage

Determining gene dosage by measuring single-cell resistance (SCR) (*Protocol 1*) also requires some kind of calibration unless only relative values are required, e.g. in the analysis of copy numbers of copy number control (*cop*) mutants.

An alternative method is to use an antibiotic resistance or other constitutive gene, (e.g. *lacZ*) and measure the specific activity of the corresponding enzyme as a measure of gene dosage and, hence, of plasmid copy number.

3.2.3 Determining copy number by measuring the rate of segregation

A third method to estimate copy numbers is based on the fact that, at least some, plasmids are randomly distributed to the daughter cells at cell division. In this case, the loss rate (*L*) is a function of the copy number at cell division (2*n*) (5,9).

$$L \approx (1/2)^{2n} \text{ per cell generation}$$

The method is further discussed in Section 4.1.

3.3 Copy (*cop*) mutants

Mutations in the plasmid genome may result in a changed copy number of the plasmid. If the plasmid carries an antibiotic resistance gene and if the mutations cause increased copy number, *cop* mutants can often be selected on plates containing higher amounts of the antibiotic than the SCR (10) (*Protocol 1*) of the wild-type. Increased antibiotic resistance can have at least four major causes:

- increased gene dosage of the resistance gene (e.g. by *cop* mutation or gene duplication)
- increased output of the drug-metabolizing enzyme (e.g. by promoter-up mutation)
- increased intrinsic resistance of the host (e.g. changes in the cell envelope decreasing the rate of uptake of the drug)
- decreased sensitivity of the target of the antibiotic

In a first screening of the mutants, the plasmid is transferred by transformation or conjugation to a new host. The SCR of the transconjugant strains is determined in order to locate the mutation to the plasmid. The size of the plasmid is then determined by gel electrophoresis, either directly or after cleavage by a restriction enzyme; a *cop* mutation, a promoter-up mutation, as well as a mutation affecting the target do not change the size of the plasmid, whereas gene duplication does. The two former classes of mutants can be distinguished by directly determining the copy number by gel electrophoresis. The relative increase in copy number can be determined as SCR or by the other methods discussed above (Section 3.2).

If the mutation causes a decrease in copy number, the mutants can not be directly selected, but the rest of the procedure described above can also be followed in this case and the effect on the copy number can be determined as SCR.

For some antibiotics, resistance increases linearly with gene dosage, whereas for others resistance is a function of gene dosage and follows a saturation curve (*Figure 3*). This fact can be used to select for conditional *cop* mutants (*Protocol 3*).

4. Partition and other maintenance functions

Plasmids are normally extremely stably maintained because they code for several systems that minimize plasmid loss at cell division in a growing population (9):

- plasmid partition systems ensure that distribution of the plasmid copies to the daughter cells is better than random (*Figure 4*)
- multimerization by recombination is counteracted by resolution systems
- killer systems kill from within those (few) cells that are born without plasmid
- cell division may be delayed if the copy number has dropped drastically

4.1 Plasmid partition

Basic replicons (the smallest part of a plasmid that is able to replicate and has the same copy number as that of the parent plasmid) are often unstably inherited, because they lack a *par* function that distributes the plasmid molecules to all daughter cells at cell division (*Figure 4*). Therefore, plasmid-free cells appear during each cell generation at a frequency that is dependent on the copy number of the plasmid. Some basic replicons have been shown to be randomly distributed at cell division. Therefore, they are lost at a frequency (*L*) that is;

$$L \approx (1/2)^{2n} \text{ per cell generation}$$

Protocol 3. Isolation of conditional *cop* mutants (30)

If judged necessary, the starting population may be mutagenized. The method is described for *ts* mutants but may easily be adjusted for the isolation of other conditional mutants, for example nonsense mutants. In the example, the plasmid mediates resistance to chloramphenicol and penicillin.

1. Grow the bacteria exponentially at 30°C to a density of 10^7 cells/ml in LB medium. Divide the culture into aliquots (only one mutant should be recovered from each aliquot).

2. Add chloramphenicol at a concentration that inhibits the growth of bacteria carrying the parent plasmid but not of those that carry a putative *cop* mutant (see *Figure 3*). Incubate for 10 min to allow chloramphenicol to act.

3. Add penicillin at a very high concentration (2 mg/ml) in order to kill all non-conditional *cop* mutants (see *Figure 3*). Continue incubation for about 60 min.

4. Harvest the cells by centrifugation. Resuspend in fresh LB medium and incubate the culture for 30–60 min at 42°C to allow for phenotypic expression.

5. Plate dilutions on different concentrations of ampicillin or benzylpenicillin to select for *cop* mutants. Incubate overnight at 40°C.

6. Test the putative mutants for their penicillin resistance at 30°C and 42°C.

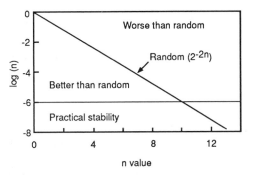

Figure 4. The frequency of formation of plasmid-free cells (*L*) as a function of the plasmid copy number (*n*) per newborn cell. The line $L = 2^{-2n}$ shows the behaviour of an ideal basic replicon. A loss rate $<10^{-6}$ per generation is experimentally difficult to determine; hence, such low loss rates are denoted 'practical stability' in the figure. The implication is that if such low loss rates are observed at a *n* value >10, it does not necessarily mean that the plasmid carries a stability function.

where $2n$ is the copy number at cell division (9). This relation is approximate since it does not take into account variations in copy number in the population of cells or the response to such variation. However, these variations do not drastically change $L = f(n)$ (5).

It has to be stressed that for some plasmids the unit of segregation is more than one genome; such plasmids segregate worse than random (see *Figure 4*) when they lack *par* and other maintenance functions (9).

Plasmid free cells (P^-) often grow somewhat more rapidly than plasmid-containing ones (P^+) (*Figure 5*). If the plasmid free derivative has a growth rate (k_2) that is higher than that of the plasmid-containing one (k_1), the loss curves are initially linear when the logarithm of the plasmid-containing fraction of the population ($P^+/(P^+ + P^-)$) is plotted against time. Then the curve gradually accelerates downwards (5) and, when the plasmid-free population becomes predominant, the slope of the curve is constant. Hence, the initial slope gives the L value and the final slope is $k_1 - k_2 - L$. An alternative way to plot the data is shown in *Figure 5*, namely $^2\log(P^-/P^+) = f(t)$. In this case, the slope of the linear part of the curve is $k_2 + L - k_1$ and the extrapolated intercept is $^2\log(L/(k_2 + L - k_1))$.

The shape of the curves in *Figure 5* is dependent upon the relative growth rates k_1 and k_2 and the L value. This is important to bear in mind when two different plasmid derivatives are compared—a more rapidly declining curve does not necessarily indicate a more rapid loss at cell division but could be due to a more drastic effect upon the growth rate of the host.

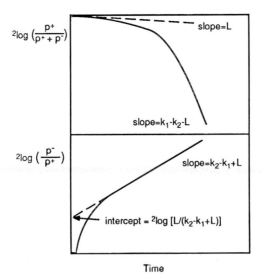

Figure 5. Kinetics of plasmid loss during exponential growth of a plasmid-containing population. Abbreviations: P^+ and P^-, number of plasmid-containing and plasmid-free cells, respectively; k_1 and k_2, growth rate (generations per hour) of the P^+ and P^- populations, respectively; L, probability of forming a plasmid-free cell at each cell division.

There are two main ways for determining loss rates (*L* values): by analysis of exponentially growing populations (*Protocol 4*), and by repeating dilutions of populations that are allowed to proceed into the stationary phase between each dilution step (*Protocol 5*). The first method gives better reproducibility and is easier to interpret, whereas the latter is easier to perform but harder to interpret, since the observed loss is dependent upon the behaviour of the plasmid during exponential growth and stationary phase.

Protocol 4. Determination of plasmid loss rates during exponential growth

1. Grow bacteria containing a plasmid with an antibiotic resistance gene exponentially in LB or minimal medium.

2. At a cell density of about 10^8/ml, dilute the population 10^4-fold in pre-warmed medium and continue incubation.

3. Repeat the dilution several times in order to keep the cell concentration between 10^4 and 10^8/ml.

4. At appropriate intervals (five to ten generation times; if the loss rate is low, the intervals should be longer), spread dilutions of the culture on antibiotic-free plates for the isolation of single-cell colonies. Incubate the plates overnight.

5. Replica plate on to plates with and without the antibiotic. Incubate the plates overnight.

6. Score the plates and calculate the fraction of plasmid-containing clones in each sample.

Protocol 5. Determination of loss rates by repeated batch cultures

1. Grow bacteria containing a plasmid with an antibiotic resistance gene overnight in LB or minimal medium.

2. Dilute the overnight (stationary) cultures 10^6-fold in the same medium and incubate overnight. Repeat dilution and overnight incubation several times. The number of repetitions depends on the degree of stability of the plasmid.

3. Spread dilutions of the overnight cultures on antibiotic-free plates. Incubate the plates overnight.

4. Replica plate on to plates with and without the antibiotic. Incubate the plates overnight.

5. Score the plates and calculate the fraction of plasmid-containing clones after each overnight incubation (20 generations).

4.2 Killer systems

There are at least two types of killer systems, *ccd* of plasmid F and *hok/sok* of plasmid R1 (9). In the case of *ccd*, the plasmids produce a killer protein and another protein that blocks the killer activity. When the plasmid is lost, the blocking agent decays but the killer protein persists and eventually kills the cells. In the case of *hok/sok*, the mRNA for the killer protein (Hok) is formed, but its expression is inhibited by an unstable antisense RNA (Sok). When the plasmid is lost, the antisense RNA decays, the extremely stable Hok mRNA is expressed and the cells are killed. Cloning of a killer function on a basic replicon stabilizes the replicon by killing the plasmid-free segregants. Methods for distinguishing between Par and Kil functions are discussed in Section 4.5.

4.3 Systems that delay cell division

The Ccd system of plasmid F was initially believed to delay cell division at very low copy numbers of F. Such a system would lead to increased stability. It is still possible that the Ccd system has an element of cell division inhibition.

4.4 Site-specific recombination systems

Site-specific recombination systems increase the stability of inheritance by resolving the dimers, trimers, etc., that are formed by homologous recombination. This increases stability by increasing the number of segregating units. However, a basic replicon carrying a site-specific recombination system can not be more stably maintained than according to random segregation (see *Figure 4*) at a copy number that equals the number of plasmid genomes per cell.

4.5 Distinction between different maintenance systems

When a system that stabilizes the inheritance of the plasmid has been identified, it has to be characterized. There are two classes of such systems, those that cause better than random segregation and those that are only able to increase the stability from worse than random to random segregation (*Figure 4*). Therefore, such an analysis requires that the copy number as well as the L value are carefully determined (9).

Site-specific recombination systems belong to the category that can increase stability only to that determined by random segregation. Similarly, putative systems that reduce the copy number distribution or that dissolve plasmid aggregates would belong to this class (9).

Three systems that give better than random segregation are true partition systems, killer systems, and systems that delay cell division in those cells in which the copy number has dropped to low values. Killer systems can be distinguished from partition systems by cloning on to a *rep(ts)* vector. If a growing population of bacteria containing such plasmids is shifted to a non-permissive temperature, plasmid-free segregants appear when the copy number of the plasmid has dropped to one per cell. Viable count curves will reach a plateau if the cloned system is a killer system, whereas they will not be affected if it is a partition system (see *Figure 6* and reference 9).

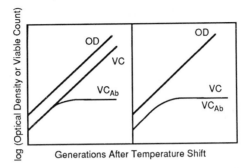

Figure 6. Effect of plasmid partition and killer functions on growth of the host bacteria after shift of an exponentially growing population to a temperature at which the *rep(ts)* plasmid (carrying a gene for resistance to the antibiotic Ab) does not replicate. Viable count is determined by incubation of the plates at permissive temperature. Abbreviations: OD, optical density; VC, viable count; VC$_{Ab}$, viable count on antibiotic-containing plates.

5. Incompatibility

Related plasmids can not be stably maintained in growing populations of bacteria; they are incompatible (11). For example this is the case for two differently labelled derivatives of the same plasmid. Incompatibility may have different causes including randomization during replication and partition; furthermore systems that stabilize plasmid maintenance by killing plasmid-free segregants from within may give rise to incompatibility effects (see below).

5.1 Incompatibility caused by randomization during replication

Plasmid replication is random and spread over the whole cell cycle as has been demonstrated by Meselson–Stahl density shift experiments (12, 13) (*Protocol 6*) (see also Section 11). Some molecules do not replicate at all during a cell cycle whereas others replicate two or more times; *Figure 7* demonstrates the fate of two differently marked derivatives of the same plasmid. Randomization takes place during replication as well as during partition and leads to distortion of the ratio between the number of molecules of each of the two derivatives in individual cells (9, 14). As these deviations can not be corrected, pure lines appear.

Protocol 6. Density shift experiments

A. *Chromosomal DNA*

1. Grow the bacteria in 25 ml of MOPS (heavy) medium with 2.0 mM [^{13}C]glucose, and 1.3 mM ^{15}NH$_4$Cl.

2. At a cell density corresponding to an A$_{450}$ of 0.5, 2 µCi [methyl-^{3}H]thymidine is added per ml of culture.

3. Incubate for 4–8 min (to pulse label).

4. Add 60 ml of light MOPS medium containing 0.67% [^{12}C]glucose, 33 mM ^{14}NH$_4$Cl, 33 µg [^{12}C]thymidine/ml, and 167 µg uridine/ml (to chase).

5. At various times after the pulse, take 10 ml samples, add 0.1 M NaN$_3$, and chill on ice.

6. Harvest the cells by centrifugation at 4°C.

7. Wash once with 5 ml of cold TES buffer (50 mM Tris–HCl, 50 mM NaCl, and 5 mM EDTA, pH 8.0) containing 100 µg thymidine/ml.

8. Suspend the bacteria in 0.1 ml of 20% sucrose, 50 mM Tris–HCl pH 8.0.

9. Add 0.1 ml of a fresh solution of lysozyme and RNase (2 and 1 mg/ml, respectively).

10. Incubate at 37°C for 30 min.

11. Add 30 µl of sodium lauryl sarcosinate (10% in 0.25 M EDTA, pH 8.0) and lyse the cells by gentle mixing.

12. Shear the lysates by mildly drawing them five times through a 21-G2 (50/8) needle.

13. Mix with 5 ml of CsCl solution. The concentration of CsCl has to be adjusted to the GC content of the DNA. For *E. coli* DNA, 1.345 g CsCl are added per ml of 10 mM Tris–HCl and 0.1 M EDTA, pH 8.0. The refractive index at 20°C is adjusted to 1.4050.

14. Separate light, hybrid, and heavy DNA by equilibrium centrifugation, e.g. in a Beckman VTi65 rotor at 30 000 r.p.m. and 15°C for 60 h.

15. Collect 40 fractions in a microtitre plate.

16. Spot 20 µl of each fraction on a strip of Whatman 3 mm paper. Dry the paper and count the samples in a scintillation counter.

B. *Plasmid DNA*

In this case, the same procedure can be used and the amount of plasmid DNA determined by hybridization of each fraction to ^{32}P-labelled plasmid-specific probes. Alternatively, plasmid DNA can be separated from chromosomal DNA in each fraction (or pools of fractions) by neutral sucrose gradient centrifugation or by gel electrophoresis. Please note that the procedure has to be adjusted if the plasmid and the chromosomal DNA have different GC contents.

Replication incompatibility alone can be studied by using two differently labelled derivatives of a basic replicon that have different *par* systems. *Figure 7* shows that randomization during replication alone leads to a lower L value than that obtained if randomization occurs both at replication and partition ((9) and *Figure 7*). The ratio is $(n + 1)/2n$, which approaches 1/2 at higher n values, i.e. both processes contribute equally to the degree of incompatibility at higher copy numbers.

Pure clones appear from the heteroplasmid population (see *Figure 7*) at a frequency that is a function of the copy number (14). Methods to quantita-

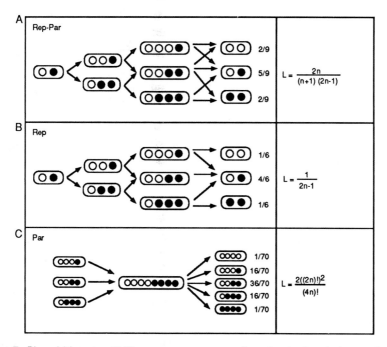

Figure 7. Plasmid incompatibility as a consequence of randomization during replication and partition. Randomization occurs during replication (B; two differently labelled derivatives (○ and ●) of a basic replicon carry different *par* systems), partition (C; two different basic replicons have different markers (○ and ●) but the same *par* system), or both (A; two differently labelled derivatives (○ and ●) of a basic replicon carry the same *par* system). The copy number at birth is denoted n. The diagrams and the mathematical expressions are simplified and statistical variations during replication and partition are ignored.

tively measure the degree of incompatibility (rate of appearance of pure clones) are shown in *Protocols 7–9*. The two first methods measure the rate of reduction of the heteroplasmid population during growth in liquid medium, whereas the last method (*Protocol 9*) determines this rate during growth from one cell to a colony (about 10^8 cells) on a plate. If cells carrying either of the two derivatives have different growth rates, the ratio between the two pure lines will deviate from 1:1. A method to determine the relative loss of each of two plasmids from a heteroplasmid population is described in *Protocol 10*.

Two derivatives of a replicon may have different efficiencies of replication, i.e. probability of being selected for replication. In a system with random selection for replication, this leads to skewed copy number distributions and,

---

Protocol 7. Quantitative incompatibility test—method 1

Two differently marked, (e.g. Km^r and Ap^r) plasmid derivatives are used.

1. Transfer one of the derivatives (by transformation, transduction, or conjugation) into a strain that already contains the other derivative.

2. Add growth medium, (e.g. LB) to the transfer mixture and incubate for about 60 min to allow phenotypic expression of resistance.

3. Spread the transfer mixture on plates containing both antibiotics.

4. Incubate overnight and suspend a colony in growth medium and incubate.

5. At intervals of e.g. 1 h over a period of 5–8 h, spread samples of the growing culture on plates with and without both antibiotics. For plasmids with higher copy numbers, the length of the sampling intervals and the duration of the experiment may have to be increased.

Either

6. Count the number of colonies and plot this number, as well as the frequency of heteroplasmid cells, as a function of time.

7. Calculate the rate of reduction of the relative size of the heteroplasmid population from the slope of the curves.

Or

6a. If the loss rate is low, an alternative is to replicate from the antibiotic-free plates on to plates with and without both antibiotics.

7a. Count the number of colonies that grow on the plates and calculate the fraction of the colonies that carry both resistances. Plot this fraction as f(t) and calculate the rate of reduction of the relative size of the heteroplasmid population from the slope of the curve.

Protocol 8. Quantitative incompatibility test—method 2

Two differently marked, (e.g. Km^r and Ap^r) plasmid derivatives are used.

1. Transfer one of the derivatives (by transformation, transduction, or conjugation) into a strain that already contains the other derivative.
2. Add growth medium (LB) to the transfer mixture and incubate for about 60 min to allow phenotypic expression of resistance.
3. Add antibiotics to counterselect the non-transformed recipient and incubate for 2 h.
4. Harvest by centrifugation, resuspend in antibiotic-free medium, and incubate.
5. At intervals of e.g. 1 h over a period of 5–8 h, spread samples of the growing culture on plates with and without both antibiotics. For plasmids with higher copy numbers, the length of the sampling intervals and the duration of the experiment may have to be increased.
6. Count the number of colonies and plot this number, as well as the frequency of heteroplasmid cells, as a function of time.
7. Calculate the rate of reduction of the relative size of the heteroplasmid population from the slope of the curves.

Protocol 9. Quantitative incompatibility test—method 3

In this method, the degree of incompatibility is measured during growth of a colony. Two differently marked, (e.g. Km^r and Ap^r) plasmid derivatives are used.

1. Transfer one of the derivatives (by transformation, transduction, or conjugation) into a strain that already contains the other derivative.
2. Add growth medium (LB) to the transfer mixture and incubate for about 60 min to allow phenotypic expression of resistance.
3. Spread the transfer mixture on plates containing both antibiotics.
4. Incubate overnight and suspend a complete colony in growth medium.
5. Spread two dilutions (α giving 200–400 and β giving 2000–4000 colonies per plate) of the transfer mixture on LA plates without antibiotics. (Assume that there are about 10^8 cells in the colony.)
6. Incubate overnight.
7. Replica plate from the α plates on to plates containing either of the antibiotics. Resuspend all cells from the β plates in medium and spread dilutions on LA plates with or without both antibiotics.

Protocol 9. *Continued*

8. Count the number of colonies. The α series gives the fraction of cells carrying both antibiotics at the time of plating and the β series the fraction of cells that still carry both plasmids after growth for about 27 generations to give a colony.

9. Calculate the rate of reduction of the relative size of the heteroplasmid population during growth from one cell to a colony.

Protocol 10. Quantitative incompatibility test—method 4

In this method, the relative loss of each of the two plasmid derivatives is determined. Two differently marked, (e.g. Km^r and Ap^r) plasmid derivatives are used.

1. Proceed as described up to step **5** in *Protocol 8*.

2. Completely resuspend at least ten colonies individually in LB growth medium. Test each population for the presence of both plasmid derivatives by spreading on plates containing either of the antibiotics.

3. Mix the cultures that contain both resistances. Dilute and spread for single-cell colonies on antibiotic-free plates. Incubate overnight.

4. Replica plate at least 500 colonies on plates without antibiotics or with either of the antibiotics. Incubate the plates overnight.

5. Count the number of colonies and calculate the relative loss of each of the plasmid derivatives.

---

hence, the ratio between the frequency of the pure lines will deviate from 1:1 (15). This distortion increases with increasing copy number due to the increasing number of selective events during the cell cycle (*Figure 8*); at very high copy numbers, virtually only one pure line is predicted to occur. This might be the explanation for the so-called Cmp (competition) phenotype of plasmids pSC101 (16) and pT181 (17).

Plasmids are present in defined copy numbers during steady-state (exponential) growth of their host bacteria. They measure their concentration (copy number) and use this measurement to adjust their replication frequency in order to compensate for deviations in copy number and thereby maintain their steady-state concentration (1). Therefore, the rate of replication per plasmid copy decreases by increasing copy number (*Figure 9*). Simple models predict an hyperbolic control function (inverse proportionality between the relative replication frequency and copy number) and some plasmids, (e.g. IncFII plasmids (18) and the ColE1 family (19)) follow such kinetics. The same molecule is used to measure plasmid concentration and to adjust the

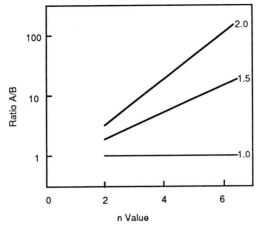

Figure 8. The ratio between the pure lines formed from a heteroplasmid population in which A and B are derivatives of the same replicon with the same *par* system. Derivative A is assumed to have a replication probability that is 2.0, 1.5, and 1.0 times higher than that of derivative B.

replication frequency; it can be a repressor protein, an antisense RNA, or DNA iterons (directly repeated sequences) (1, 18). The copy number control genes are denoted *cop*.

Figure 9 predicts that the replication frequency of a plasmid would be reduced if an active *cop* gene were cloned on to a compatible plasmid and introduced into a cell containing the parent plasmid; the cloned gene would cause incompatibility. This can be demonstrated as shown in *Protocol 11*. The loss rate caused by the cloned gene may sometimes be small. In such cases, the degree of incompatibility can be tested as in *Protocol 9* and the polarity of

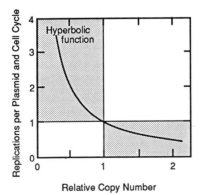

Figure 9. Kinetics of control of replication. The control curve lies within the hatched areas. Some plasmids, (e.g. plasmid R1) follow the hyperbolic ($y = k/x$) curve shown.

19

Protocol 11. Test for incompatibility effects of cloned genes

1. A vector carrying a cloned gene is transferred by transformation into a strain that contains the mother plasmid of the cloned gene.

2. Add LB growth medium to the transfer mixture and incubate for about 60 min to allow phenotypic expression of the resistance gene (A) present on the cloning vector. The resident plasmid mediates resistance to antibiotic B.

3. Spread on LA plates that select for the vector by containing antibiotic A. Incubate overnight.

4. Restreak one colony on plates containing antibiotic A. Incubate overnight.

5. Test for the presence of the parent plasmid by replica plating 50–100 colonies on plates with and without antibiotic B. Incubate the plates overnight.

6. Count number of colonies on both types of plates and calculate the frequency of loss of the resident plasmid. Normally, an *inc* gene causes a loss of the resident plasmid in more than 98% of the clones.

---

the effect as in *Protocol 10*. A cloned *cop* gene may reduce the copy number without causing loss of the parent plasmid (e.g. the *copB* gene of plasmid R1 (20)). Such effects may be discovered by measuring single-cell resistance to ampicillin (if a *bla* (β-lactamase) gene is carried by the parent plasmid) in the presence of selection for the cloning vector (20).

The use of incompatibility to identify *cop* genes is discussed in Section 9.1.

5.2 Incompatibility caused by the partition process

Incompatibility may have causes other than randomization during replication (9, 11, 21). *Figure 7* demonstrates that two different (compatible) replicons that have the same *par* system are incompatible. In a heteroplasmid population, pure lines appear at a frequency that is dependent upon the copy number. Partition incompatibility is weaker than replication incompatibility, since the copy number of each of the two plasmids is unaffected by the presence of the other plasmid. However, both replication and partition cause randomization and contribute about equally to the degree of incompatibility observed when two replicons with the same partition system are introduced into the same cell (see Section 5.1).

Also in the case of *par*, a cloned *par* gene causes incompatibility. The loss rate increases with increasing copy number of the cloning vector. However, the loss rate can never exceed that of a Par⁻ derivative (9).

5.3 Incompatibility caused by killer systems

If a basic replicon carries genetic information for a killer system, it appears to be stably maintained because the plasmid-free progeny are killed from within. However, if the same killer system is cloned on to a vector, cells lacking the basic replicon survive, and the killer system appears to express incompatibility (9).

5.4 Comparison between different incompatibility genes

A genetic analysis of a plasmid (*Protocol 11*) may reveal one or more *inc* genes. How to relate these to replication, partition, or killer systems is discussed in Sections 4.5 and 9.1.

6. Basic replicons

Natural plasmids often consist of several replicons, some of which may be dormant. The replicon(s) can be identified and characterized by:

- deletion analysis
- subcloning
- construction of cointegrates
- incompatibility testing
- probing with replicon-specific probes

In addition, the analysis of replicons often starts with determination of the nucleotide sequence of the plasmid.

6.1 Identification of replicons by deletion analysis

Natural plasmids often carry antibiotic resistance genes. The plasmid is digested (totally or partially) with a restriction endonuclease, religated, and transformed into a recipient strain. The process is repeated with other restriction endonucleases until the smallest replicon has been identified. This can be further trimmed down in size by, for example, use of the exonuclease *Bal*31. Standard methods (8) are used in this process.

6.2 Identification of replicons by cloning

This is a modification of the method described in Section 6.1. The plasmid is cleaved with restriction endonucleases and a non-replicating fragment containing a structural gene for resistance to an antibiotic is used to supply a selectable marker.

6.3 Identification by the formation of cointegrates

A negative result in the methods described in Sections 6.1 and 6.2 is difficult to interpret. Therefore, the search for replicons often starts with the construction of a cointegrate between the fragments and a conditional replicon, e.g. a *rep*(*ts*) vector or one belonging to the ColE1 or p15 families. The latter are dependent upon DNA polymerase I for their replication. In a *polA*(*ts*) host, survival of the cointegrate at a non-permissive temperature indicates that the cloned replicon is active. Further trimming can be performed as described above (Section 6.1). The deletion derivatives are tested for replication proficiency by temperature shifts in the *polA*(*ts*) host.

6.4 Identification of replicons by incompatibility typing

Plasmids belonging to the same replicon type can not stably coexist in the same cells; they are incompatible (11). This can be used to classify replicons. The plasmid is introduced into a set of strains each carrying a plasmid belonging to a known incompatibility group (22) (see *Protocol 7*).

6.5 Identification of replicons by replicon-specific probes

Replicons can be identified by hybridization to replicon-specific probes (23). This method has several advantages:

(a) It identifies a replicon even if it is compatible with the test replicon due to mutations that change the specificity of the copy number control.

(b) It is able to establish the existence of several replicons in a plasmid, which is often the case for large plasmids.

(c) It recognizes non-active, dormant, or incomplete replicons.

6.6 Mode of replication

Plasmids may replicate in at least three different ways:

(a) By theta replication, initiated from origins of replication and requiring DNA polymerase III; this replication can be uni- or bidirectional.

(b) By rolling circle replication from a nick in one of the strands. In this case, single-stranded circles are intermediates. This is typical of many small, multicopy-number plasmids in Gram-positive bacteria.

(c) By displacement replication which starts from an RNA primer and first uses DNA polymerase I. After replication of about 500 bp, normal replication catalysed by DNA polymerase III is initiated. This applies to the ColE1 and p15 families.

7. Origins (ori) of replication

Plasmids contain one or more active origins of replication (*Table 1*).

7.1 Identification of origins

Origins can be identified in several ways:

- by cloning
- by restriction analysis of replicative intermediates
- by electron microscopy
- by two-dimensional gel electrophoresis

7.1.1 Cloning of origins

Origins may be identified by ligating fragments from a plasmid with a fragment that codes for an antibiotic resistance determinant. The mixture is introduced into a host that already carries the gene for the Rep protein. Plating in the presence of the appropriate antibiotic selects for replication from the origin. The *ori* fragment can then be trimmed down in size as discussed in Sections 6.1 and 6.2

7.1.2 Identification by restriction analysis of replicative intermediates

This requires synchronization of initiation of replication, which can be achieved *in vivo* by inhibiting protein synthesis to block initiation; upon release of the block, synchronous initiation occurs. Synchrony is easiest to obtain *in vitro*. Both *in vivo* and *in vitro*, replication is initiated in the presence of radioactively labelled trinucleotides. Samples are taken at intervals, cleaved by restriction endonuclease, and analysed by gel electrophoresis. Radioactivity first appears in the fragment that is replicated first, the *ori* fragment, and then, gradually, more and more radioactive fragments appear. *In vitro*, the method can be improved by the addition of chain terminating dideoxynucleotides.

7.1.3 Identification by electron microscopy

This method requires that there are well separated single sites for cleavage by at least two different restriction endonucleases. Replicative intermediates are isolated and cleaved separately by these enzymes. In the electron microscope, an origin is shown as a bubble with two tails. The length of these tails and of the bubble is measured. The length of the tails is plotted as a function of the size of the bubble (= the length replicated) (*Figure 10*). If replication is unidirectional, the length of one arm is initially constant, whereas that of the other decreases as replication proceeds. On the other hand, if replication is bidirectional, the length of both arms decrease from the very onset of replica-

23

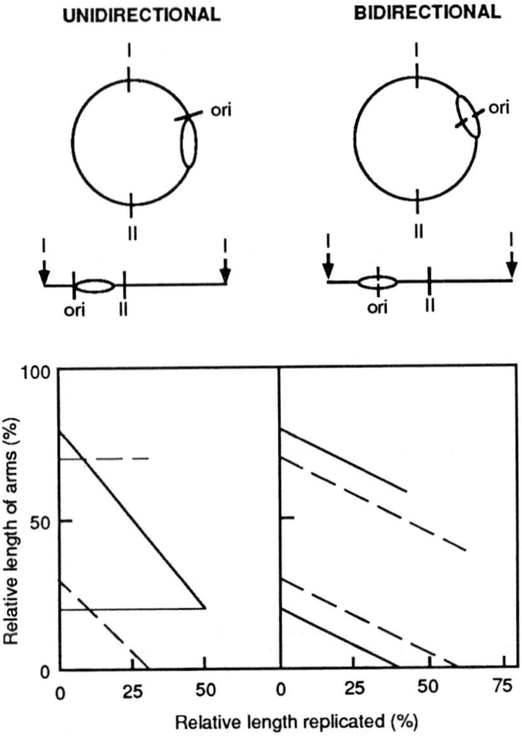

Figure 10. Determination of the location of *ori* by EM. Replicative intermediates are collected (24), cut with restriction endonucleases I or II, and studied by EM. The length of the arms is plotted as a function of the fraction of the plasmid that has replicated (the length of the bubble). *Left*: unidirectional replication. *Right*: bidirectional replication. The results with restriction endonuclease I and II are shown as *solid* and *broken lines*, respectively.

tion. The extrapolated arm lengths (length at zero time) give a measure of the distance between the restriction site and *ori*. The location of *ori* can be calculated from these measurements. The map position of *ori* is unambiguously deduced from the result of experiments with two different enzymes (24).

7.1.4 Identification of origins by two-dimensional gel electrophoresis

In two-dimensional (2D) gel electrophoresis the DNA is separated by size in one orientation and by size and shape in the other (25, 26). *Figure 11* shows the structure of the restriction fragments obtained by cleavage of replicative intermediates. Five types of fragments appear: the linear (unreplicated) fragment, the fragment with a bubble of increasing length, the Y structure of increasing molecular weight, the double-Y of increasing molecular weight, and the fully replicated, non-separated dimer. The middle three of these

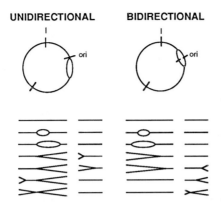

Figure 11. The type of structures found after cleavage of replication intermediates of a plasmid with restriction endonuclease I. The structure of two fragments (the origin fragment and an outside fragment) as they appear when replication proceeds are shown in the *lower part* of the figure.

types of molecules give distinct and differently shaped curves in 2D gel electrophoresis whereas the first and the fifth molecules give distinct spots. Hence, hybridization with specific probes gives distinct 2D patterns that allow the identification of how different fragments replicate:

- bidirectionally from an *ori* within the fragment
- unidirectionally from an *ori* within the fragment
- by the passing of a replisome from left or right or from right to left
- by two replisomes that meet within the fragment

In addition, this method gives information about regions where replication slows down and about termination sites (27).

7.2 Directionality of replication

7.2.1 Determination of directionality by electron microscopy

The method discussed in Section 7.1.3. also gives information about whether replication is uni- or bidirectional.

7.2.2 Determination of directionality by 2D gel electrophoresis

The methods mentioned in Section 7.1.4 also allows determination of whether replication is uni- or bidirectional.

7.3 Initiation of unidirectional replication

Some plasmids replicate unidirectionally. This type of replication leaves 3' and 5' ends at the site of initiation of replication. The 5' end can be determined by primer extension (28), whereas the 3' end can be identified after cleavage of the DNA by restriction endonucleases followed by gel electrophoresis under denaturing conditions; probing with radioactive probe will identify the 3' end fragment as a piece of single-stranded DNA that is shorter than the full-length fragment.

7.4 Rolling circle replication

This type of replication starts with a nick in one of the DNA strands; the plasmid-encoded Rep protein carries a nicking–closing activity (29). The nick site can be identified *in vitro*, since the Rep protein has nicking–closing activity which leads to relaxation of supercoiled plasmid DNA. After treatment of the relaxed DNA with proteolytic enzymes, the DNA is cleaved by a restriction endonuclease and the nick in the DNA can be identified as two specific single-stranded pieces of DNA. The exact location of the nick can be determined by measurement of the size of the two fragments and by determining their sequences.

The rolling circle type of replication is initiated from the nick site. Single-stranded circles are formed as intermediates. They can be identified by gel electrophoresis under non-denaturing conditions (30). The circularity of the single-stranded material can be visualized by electron microscopy.

8. Rep functions

Most plasmids (with the exception of the ColE1 and p15 families) code for at least one Rep protein (reference 1, and *Table 1*).

8.1 Identification

The existence of a *rep* gene is often indicated by genetic evidence. Conditional mutants (*ts* or *amber*) are discovered by replica plating of a population carrying the plasmid on to plates, selecting for the presence of the plasmid. If the plasmid is conditional for replication, no colonies are obtained at non-permissive conditions (31). The population is grown in the presence of [^{3}H]thymidine or [^{3}H]thymine at the permissive temperature and then shifted to non-permissive temperature. Samples are taken at intervals and plasmid DNA is separated from chromosomal DNA by gel electrophoresis or by ultracentrifugation (CsCl/ethidium bromide (*Protocol 2*) or neutral sucrose); the incorporation of radioactivity into plasmid DNA can then be determined.

The character of the mutation as one affecting a *rep* gene is revealed from sequence analysis of the mutant and the wild-type plasmid. From the sequence, gene fusions can be constructed between the putative *rep* gene and a reporter gene. This will allow a test of whether or not the *rep* gene is active (32).

If a *rep* gene has been cloned together with its promoter and ribosome-binding site it can be introduced into a cell carrying the origin sequence from the same plasmid to demonstrate the activity of the *rep* gene. In those cases

26

where an *in vitro* system has been developed, this test may be performed *in vitro* (*33*).

8.2 *Cis/trans activity*

Cloned *cop* genes always cause incompatibility. In control by antisense RNA, *cop* mutations sometimes lead to changes in the specificity such that the mutant antisense RNA is no longer able to interact with the wild-type target and vice versa. This is illustrated in *Figure 12*, which shows the properties of the antisense RNA (CopA) and its target (CopT) for wild-type and *cop* mutants of IncFII plasmids (e.g. R1). The primary interaction between these two RNAs occurs between two complementary loop structures and presumably involves three nucleotides in each RNA as shown in the figure. A single base pair substitution in mutant *cop-1* leads to loss of the ability of the wild-type CopA to interact with the mutant target, whereas the mutant CopA is able to interact with the wild-type target. For the mutant *cop-2*, both heterologous pairs are unable to interact. In some cases, this results in total loss of incompatibility, i.e. the mutant and the wild-type plasmid can stably coexist in a cell line (11). This is due to the fact that the Rep protein is *cis*-acting, i.e. it preferentially replicates the origin that is present on the same DNA molecule as the corresponding mRNA was transcribed from (11, 34, 35) (*Figure 13*). In other cases, the Rep protein functions equally well in *trans* and in *cis*; the mutant and wild-type plasmids share the pool of Rep molecules (*Figure 13*). In these cases, the copy number of the wild-type tends to increase and that of the mutant to decrease compared to the situation when they are alone (36).

9. Cop functions

Plasmids measure their copy numbers and use this to adjust their replication frequency (*Figure 9*). This is achieved by *cop* genes.

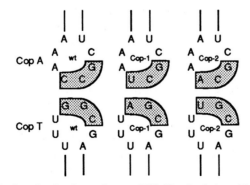

Figure 12. Control of replication by antisense RNA (CopA of plasmid R1) that binds to its complementary target (CopT). The primary interaction presumably involves three nucleotides (*boxed*) in each of the two reactants.

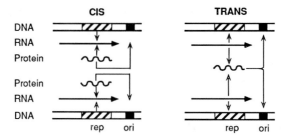

Figure 13. Control of replication by regulation of the synthesis of a rate limiting Rep protein that acts only in *cis* (*left*) or in *cis* as well as in *trans* (*right*).

9.1 Identification of *cop* genes by cloning

A *cop* gene can be identified by the cloning of fragments of a plasmid and testing them as described in *Protocol 11*. By using different restriction endonucleases and further deletion analysis as outlined in Sections 6.1 and 6.2, the minimal *inc* fragment can be defined. It should again be stressed that incompatibility may be caused by systems other than Cop. It is, therefore, necessary to determine whether the *inc* fragment is involved in copy number control.

The location of the *inc* gene in the vicinity of *ori* is an indication, but not a proof, of its involvement in copy number control.

Strong incompatibility (*Protocol 11*) is characteristic of *cop* functions. However, it should be stressed that *par* genes may also express strong incompatibility against low copy number plasmids (see *Figure 7*).

In most cases, Par incompatibility is much less severe than Cop incompatibility. This may be used to distinguish between *cop* and *par* genes. To test this, heteroplasmid populations containing the cloned gene and the parent plasmid are constructed. The SCR (*Protocol 1*) for an antibiotic to which the parent plasmid mediates resistance is then determined in the presence of selection for the cloning vector.

A cloned *cop* gene may reduce the copy number without causing loss of the parent plasmid (e.g. the *copB* gene of plasmid R1), which may be discovered as described in the previous paragraph (20).

For low copy number plasmids, a more direct technique may be used to measure effects of a *cop* gene on replication. The plasmid is introduced into a *dnaA*(*ts*) host and temperature-resistant clones are isolated. These are formed by spontaneous integration of the plasmid into the chromosome, whose replication becomes controlled by the plasmid. The cloned *cop* gene is introduced into the *int* strain, which is grown at the permissive temperature and then shifted to non-permissive temperature. DNA replication is measured during the post-shift period. A *cop* gene results in switch-off of replication (37).

In many cases, the *cop* gene acts by inhibiting the synthesis of a rate limiting Rep protein. This can be tested by the construction of a fusion between the *rep* gene and a reporter gene, (e.g. *lacZ*). The synthesis of the hybrid protein is then determined in the absence and presence of the cloned *cop* gene (32).

A comparison between the nucleotide sequence of the wild-type *inc* gene and that of *cop* mutants will supply information about the function of the cloned gene.

9.2 Characterization of *cop* genes

The Cop functions can be either repressor proteins, antisense RNAs, or DNA iterons (1). Once a *cop* gene has been identified, it has to be characterized.

The nucleotide sequence of the minimal *cop* fragment is searched for ORFs (open reading frames), promoters, ribosome-binding sites, iterons (inverted and direct), and other landmarks. This may give clues about the Cop function. Further characterization requires a detailed analysis of the sequence, site-directed mutagenesis, and functional tests. It is not possible to propose a general scheme for this analysis.

9.3 Runaway replication

The copy number of plasmids that control their replication by antisense RNA is sensitive to changes in the relative strength of the promoters from which sense and antisense RNA are transcribed. The two promoters are located in the same region of the plasmid and the transcripts meet in between the promoters. Even moderate increases in the strength of the sense promoter may drastically reduce transcription from the antisense RNA promoter. This causes the copy number to increase drastically by increasing the strength of the sense promoter and may even lead to a loss of copy number control, so-called runaway replication. This is very useful because not only plasmid DNA is amplified, but also RNA and protein formed from this DNA. Runaway replication vectors are therefore useful as production vectors (38).

9.4 Kinetics of copy number control

Figure 9 shows that the rate of replication per plasmid copy decreases with increasing copy number. The actual kinetics can be analysed as follows:

- analysis of plasmid replication after shift from copy numbers that have been artificially increased by cointegrate formation
- analysis of plasmid replication during metabolic shifts
- analysis of expression of *rep* genes at different concentrations of *cop* genes

9.4.1 Plasmid replication during shift from artificially increased copy numbers

Cointegrates between active replicons often have the same copy number as that of the replicon that has the highest copy number. This was first demonstrated by Cabello *et al.* (39) for a hybrid between plasmids ColE1 and pSC101. Electron microscopy showed that only the ColE1 origin was used and that the pSC101 replicon was totally switched-off when the copy number was increased four-fold by cointegrate formation. In other cases, the hybrid shows a lower copy number, e.g. in hybrids between ColE1 and plasmid R1 (40); this is not understood. However, this fact was used to select for recombination between the chromosome and plasmid pBR322 carrying a derivative of plasmid R1 flanked by chromosomal segments. The copy number of such a hybrid is much lower than that of pBR322. Upon homologous recombination between the chromosomal segments on the plasmid and on the chromosome, plasmid R1 is integrated into the chromosome whereas pBR322 is released. This causes the copy number of the plasmid to become normal (i.e. high). Hence ampicillin resistance mediated by the *bla* gene located on pBR322 increases about ten-fold, which offers a strong selection pressure for the isolation of recombinants (40).

Tsutsui and Matsubara (41) constructed a cointegrate between plasmids pBR322 and F. The copy number of F was thereby greatly increased. The host was PolA(ts). By shifting an exponentially growing population from a permissive to a non-permissive temperature, replication of pBR322 ceased and the copy number of the cointegrate decreased as the population of bacteria increased. By pulse labelling the DNA, the rate of plasmid replication was measured at different times after the shift. Plasmid F did not replicate until the copy number was less than two-fold higher than that of the free plasmid. Hence, the control curve is close to a switch function (1). In a similar experiment (42), a high copy number and Rep(ts) derivative of plasmid F was fused to plasmid R1. The rate of R1 replication was followed after shifting to a non-permissive temperature. In this case, there was no switch-off. The control curve was found to be hyperbolic (see *Figure 9*). Since the probability of replication of each plasmid copy is inversely proportional to the copy number, the replication frequency per cell is constant and independent of the copy number. This is referred to as the $+n$ mode of replication (43).

It is not possible to construct a standard protocol for this type of analysis, since the properties of each cointegrate has to be determined and the conditions that are non-permissive for replication of the high copy number moiety may vary.

9.4.2 Analysis of plasmid replication during metabolic shifts

Plasmids often have different copy numbers at different growth rates. This allows measurement of plasmid replication during shifts between different copy numbers (44). However, metabolic shifts may cause drastic physiological disturbances. One way to reduce this disruption is to perform growth rate shifts using glucose minimal medium with and without α-methylglucose (*Protocol 12*); this substance is not metabolized by *E. coli* but is a competitive inhibitor of the uptake of glucose (45).

Protocol 12. Plasmid replication during copy number shifts (44)

The protocol includes an upshift and a downshift.

1. Grow the bacteria exponentially in minimal medium containing glucose (0.2%), adenosine (250 μg/ml), and [^{14}C]thymidine (0.25 μCi/ml, specific activity 50–100 mCi/mmol) (culture A) and, in addition, α-methylglucose (αMG, 2%)(culture B) to a cell density of about 10^8/ml.

2. Take a sample from each culture for the determination of plasmid DNA.

3. Add αMG (2%) to culture A and glucose (1.8%) to culture B.

4. Divide both cultures into aliquots and continue incubation.

5. At intervals add [^{3}H]thymidine (5 mCi/ml, specific activity about 20 Ci/mmol) to one culture from each set. Continue incubation for 20 min (culture A) and 10 min (culture B).

6. Add cold thymidine (100 μg/ml) to chase the radioactivity. Continue incubation for another 5 min.

7. Harvest the cells after addition of an equal volume of cold TES buffer. Collect the cells and determine the amount of ^{14}C and ^{3}H radioactivity in plasmid and chromosomal DNA by CsCl/ethidium bromide gradient centrifugation (*Protocol 2*).

9.4.3 Analysis of expression of *rep* genes at different concentrations of *cop* genes

Plasmid replication, at least in many cases, is controlled at the level of expression of a rate limiting Rep protein. Hence, the kinetics of control of replication can be studied indirectly. Fusions between the *rep* gene and a reporter gene (e.g. *lacZ*) are constructed and the specific activity of the hybrid protein (e.g. β-galactosidase) is measured in the presence of the *cop* gene (cloned on to vectors with different copy numbers) from the same plasmid (46). A general protocol can not easily be devised.

10. Replicative intermediates

Plasmids may replicate in at least three different ways. Replication intermediates have to be isolated if origins and the directionality of replication are to be analysed, e.g. by electron microscopy. This isolation can be achieved by ultracentrifugation in sucrose gradients (47). In neutral sucrose gradient centrifugation, two main plasmid peaks appear, corresponding to open circular (OC) and covalently closed circular (CCC) DNA. However, replicative intermediates appear as material that sediments more rapidly. In alkaline sucrose gradients, replicative intermediates appear between the OC and CCC peaks. In many respects, gel electrophoresis resembles velocity gradient centrifugation. Hence, replicative intermediates can also be separated on gels.

11. Timing of replication

Plasmid replication is independent of that of the chromosome and is generally believed to take place with equal probability during the entire cell cycle. This can be analysed in at least three different ways:

- in the baby machine
- by density shift experiments
- by analysis of DNA replication in cells of different size

11.1 The baby machine

The bacteria are collected on a filter. The filter is then inverted and pre-warmed growth medium is poured through it. One of the daughter cells formed at each cell division is spontaneously released and can be collected. Since the daughter cells are first formed from the oldest cells on the filter and then from progressively younger ones, the baby machine can be used to study events in the cell cycle. If the DNA of the population is pulse labelled just before filtering, the radioactivity of the samples gives a measure of DNA replication as a function of cell age. This can be related to the cell cycle itself. The procedure has been described in many papers by Helmstetter and Cooper, e.g. (4), (48), (49)

11.2 Meselson–Stahl density shift experiments (*Protocol 6*)

The principle of studying replication of chromosomal DNA by density shift experiments is shown in *Figure 14* (12). A culture is grown exponentially in dense (^{13}C and ^{15}N) medium and is pulse labelled with [^{3}H]thymidine or [^{3}H]thymine. The radioactivity is chased with an excess of cold thymidine or thymine in light (^{12}C and ^{14}N) medium. Samples are taken at intervals and heavy, hybrid, and light plasmid DNA are separated by CsCl/ethidium bromide gradient centrifugation. The relative amount of radioactivity in heavy and hybrid DNA is determined. In this method, the timing of the next round of replication of the DNA that was labelled before the density shift is determined.

The same procedure as described in the previous paragraph can be used in studies of plasmid replication; the amount of plasmid DNA is determined by hybridization of each fraction to ^{32}P-labelled plasmid-specific probes. Alternatively, plasmid DNA can be separated from chromosomal DNA in each fraction (or pools of fractions) by neutral sucrose gradient centrifugation (50) or by gel electrophoresis. Please note that the procedure has to be adjusted if the plasmid and the chromosomal DNA have different GC contents.

11.3 Analysis of DNA replication in cells of different size

Exponentially growing bacteria are pulse labelled with [^{3}H]thymidine or [^{3}H]thymine and harvested. The cells are separated by size on a gradient and

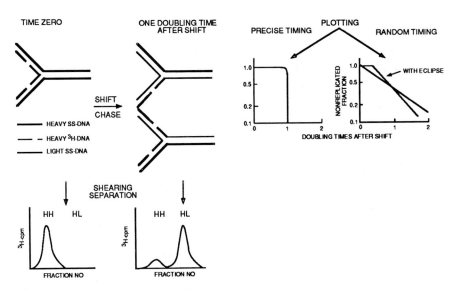

Figure 14. Principle of Meselson–Stahl density shift experiment.

the radioactivity of the plasmid and chromosomal DNA is determined in each cell fraction (50) (*Protocol 13*).

11.4 What do the methods actually measure?

The three methods described in this section do not measure the same thing. Analysis of cells separated by size allows the location of plasmid and chromosome replication to cell size (age). However, separation is not very clean and the fractions overlap. Baby machine experiments give a somewhat better cell age resolution but there are some problems with defining the timing of cell division. Finally, density shift experiments do not directly relate replication to the cell cycle but give information about interreplication times, i.e. the time when a second replication event occurs in DNA that was labelled during the pulse. There is an eclipse immediately after one replication during which a second replication cannot occur (51). Cell cycle-specific replication is difficult to distinguish from random replication with a long eclipse period.

12. *In vitro* replication

In vitro replication systems have been developed for several different plasmids (*Table 2*). Such systems have to be developed for each plasmid and no general protocol can be recommended. Some of the systems are not well defined and only crude extracts are used. The different systems are discussed in reference 52.

Protocol 13. Plasmid replication in cells of different size (50)

1. Grow the bacteria exponentially in 20 ml minimal medium containing [^{14}C]thymidine (about 0.2 μCi/ml; specific activity 50–100 mCi/mmol).

2. At a cell density of about 10^8/ml, add [^{3}H]thymidine (about 5 μCi/ml; specific activity about 25 Ci/mmol).

3. After about 1/10th of a generation time, chill the culture rapidly and wash it twice with ice-cold medium containing 0.02% formalin.

4. Suspend the pellet in 1 ml of medium and layer it on to a sucrose step gradient (obtained by layering 25 ml each of 12, 25, and 33% sucrose solutions in a centrifuge tube).

5. Run the gradients at 1000 r.p.m. for about 45 min in a 6 × 100 ml swing-out rotor.

6. Collect about 100 fractions of the cell band.

7. Precipitate every third fraction in the cell band with 10% TCA (trichloroacetic acid) and measure the ^{14}C activity. Plot the cumulative ^{3}H counts as a function of the cumulative ^{14}C counts—this gives an estimate of the lengths of the B, C, and D periods in the cell cycle.

8. Pool the remaining fractions into six portions, all containing about the same amount of ^{14}C activity, such that the first portion contains the largest cells, and so forth, until the sixth portion, that contains the smallest cells.

9. Pellet the cells in each portion, wash with TES (50 mM Tris–HCl, 50 mM NaCl, 5 mM EDTA, pH 8.0) buffer. Resuspend in 1 ml TES buffer. Add 0.2 ml of freshly prepared lysozyme solution (10 mg/ml lysozyme, 0.1 mg/ml RNase). Incubate for 5 min on ice. Add 0.4 ml of EDTA (0.25 M, pH 8.0). Incubate for 5 min on ice. Add 1.6 ml of lysis buffer (50 mM Tris–HCl, 63 mM EDTA, 1% Brij-58, 0.4% sodium deoxycholate, pH 8.0). Incubate on ice for about 5 min or until the solution becomes viscous.

10. Separate plasmid and chromosomal DNA by centrifugation in a CsCl/ethidium bromide gradient (*Protocol 2*). Collect about 40 fractions. Measure the ^{3}H and ^{14}C radioactivity in each fraction in a liquid scintillation counter. Calculate the total radioactivity in the plasmid and chromosomal peaks. The ratio between the ^{3}H and ^{14}C counts gives a measure of the relative rate of plasmid and chromosome replication in each of the six portions analysed.

Alternative procedure

7. Run the fractions on a gel and measure the radioactivity in the plasmid and chromosomal bands in a PhosphorImager (Molecular Dynamics, Sunnyvale CA).

Table 2. *In vitro* replication systems

Plasmid	Reference
R6K	Inuzuki and Helinski (54)
ColE1	Itoh and Tomizawa (55)
R1	Diaz *et al.* (56)
F	Muraiso *et al.* (57)

13. Visualization of plasmids

Plasmids can be visualized by at least two different methods:

- electron microscopy
- fluorescence microscopy

13.1 EM

DNA can be visualized in the electron microscope. This can be used in many types of studies of plasmids:

(a) Size (molecular weight) as contour length; 1 μm corresponds to about 2.0 × 10⁶ in molecular weight or 3.0 kb.

(b) Homology between DNA by heteroduplex mapping.

(c) Structure of DNA, e.g. topology (linearity, circularity, supercoiling) or the presence of inverted segments, IS sequences, and transposons.

(d) Localization of replicational origins (see Section 7.1.3) and determination of directionality of replication (see Section 7.2.2).

(e) Localization of sites for binding of specific proteins to DNA, e.g. RNA polymerase to promoters.

These methods are briefly discussed in reference (52) and in papers referred to there.

13.2 Fluorescence microscopy

Plasmids can be visualized in cells after staining with DAPI as described in *Protocol 14* (53). Plasmids are easier to distinguish from chromosomal DNA in chromosome-less minicells or in cells in which growth has been allowed to continue in the absence of chromosome replication.

Acknowledgements

Our work has been supported by the Swedish Cancer Society, the Swedish Natural Science Research Council, and The Knut and Alice Wallenberg Foundation.

Protocol 14. Visualization of plasmids by fluorescence microscopy

Plasmids are easier to observe if the amount of chromosomal DNA is decreased, e.g. by growing a *dna*(ts) mutant at a non-permissive temperature for one or two doubling times to inhibit chromosome replication or by using a minicell-producing strain.

1. Smear the bacteria on to a microscope slide, allow them to dry, thoroughly fix them by the addition of two or three drops of cold (−20°C) methanol, and allow them to dry again. Repeat the fixation procedure three times. Wash the slides four to six times with large volumes of tap water. Let the slides dry at room temperature.

2. Smear 10 μl of poly-L-lysine (5 μg/ml) over the dried cells and let the samples dry.

3. Add about 10 μl DAPI (4′,6′-diamidino-2-phenylindole) solution (0.2 μg/ml in 40% glycerol).

4. Analyse the cells in a fluorescence microscope, e.g. an Olympus model BHS phase-contrast microscope supplied with a reflected light fluorescence attachment and a DPlan apochromatic ×100 UVPL objective.

References

1. Nordström, K. (1985). In *Plasmids in bacteria* (ed. D. R. Helinski, S. N. Cohen, D. B. Clewell, D. A. Jackson, and A. Hollaender), pp. 189–214. Plenum Publ. Co., New York.
2. Uhlin, B. E. and Nordström, K. (1977). *Plasmid*, **7**, 1.
3. Lundbäck, A. and Nordström, K. (1974). *Antimicrob. Ag. Chemother.*, **5**, 492.
4. Cooper, S. and Helmstetter, C. E. (1968). *J. Mol. Biol.*, **31**, 519.
5. Nordström, K., Molin, S., and Aagaard-Hansen, H. (1980). *Plasmid*, **4**, 215.
6. Collins, J. and Pritchard, R. H. (1973). *J. Mol. Biol.*, **78**, 143.
7. Clewell, D. B. and Helinski, D. R. (1969). *Proc. Natl. Acad. Sci. USA*, **64**, 599.
8. Maniatis, T., Fritsch, E. F., and Sambrook, J. (ed.) (1982). *Molecular cloning, a laboratory manual*. Cold Spring Harbor Laboratory Press, NY.
9. Nordström, K. and Austin, S. (1989). *Annu. Rev. Genet.*, **23**, 37.
10. Nordström, K., Ingram, L. C., and Lundbäck, A. (1972). *J. Bacteriol.*, **110**, 562.
11. Novick, R. P. (1987). *Microbiol. Rev.*, **51**, 381.
12. Meselson, M. and Stahl, F. (1958). *Proc. Natl. Acad. Sci. USA*, **44**, 671.
13. Gustafson, P. and Nordström, K. (1975). *J. Bacteriol.*, **123**, 443.
14. Nordström, K., Molin, S., and Aagaard-Hansen, H. (1980). *Plasmid*, **4**, 332.
15. Nordström, K., Molin, S., and Aagaard-Hansen, H. (1981). In *Molecular biology, pathogenicity, and ecology of bacterial plasmids* (ed. R. C. Clowes, S. B. Levy, and E. L. Koenig), pp. 291–301. Plenum Publ. Co., New York.

16. Tucker, W. T., Miller, C. A., and Cohen, S. N. (1984). *Cell*, **33**, 191.
17. Gennaro, M. L. and Novick, R. P. (1986). *J. Bacteriol.*, **168**, 160.
18. Nordström, K., Molin, S., and Light, J. (1984). *Plasmid*. **12**, 71.
19. Bremer, H. and Lin-Chao, S. (1986). *J. Theor. Biol.*, **123**, 453.
20. Nordström, M. and Nordström, K. (1985). *Plasmid*, **13**, 81.
21. Austin, S. A. and Nordström, K. (1990). *Cell*, **60**, 365.
22. Novick, R. P., Clowes, R. C., Cohen, S. N., Curtiss III, R., Datta, N., and Falkow, S. (1976). *Bacteriol. Rev.*, **40**, 168.
23. Coutourier, M., Bex, M., Bergquist, P. L., and Maas, W. K. (1988). *Microbiol. Rev.*, **52**, 375.
24. Timmis, K. N., Cohen, S. N., and Cabello, F. C. (1978). In *Prog. Mol. Subcell. Biol.* (ed. F. E. Hahn), pp. 1–58. Springer-Verlag, New York.
25. Brewer, B. J. and Fangman, W. L. (1987). *Cell*, **51**, 46.
26. Huberman, J. A., Spohla, L. D., Nawotki, K. A., El-Assouli, S. M., and Davis, L. R. (1987). *Cell*, **51**, 473.
27. Greenfelder, S. A. and Newlon, C. S. (1992). *Mol. Cell. Biol.*, **12**, 4056.
28. Bernander, R., Krabbe, M., and Nordström, K. (1992). *EMBO J.*, **11**, 4481.
29. Koepsel, R. R., Murray, R. W., Rosenblum, W. D., and Khan, S. A. (1985). *Proc. Natl. Acad. Sci. USA*, **82**, 6845.
30. te Riele, H., Michel, B., and Ehrlich, S. D. (1986). *Proc. Natl. Acad. Sci. USA*, **83**, 2541.
31. Gustafsson, P. and Nordström, K. (1978). *Plasmid*, **1**, 134.
32. Light, J. and Molin, S. (1981). *Mol. Gen. Genet.*, **184**, 56.
33. Masai, H., Kaziro, Y., and Arai, K.-I. (1983). *Proc. Natl. Acad. Sci. USA*, **80**, 6184.
34. Masai, H. and Arai, K.-I. (1988). *Nucleic Acids Res.*, **84**, 4781.
35. Easton, A. M. and Rownd, R. H. (1982). *J. Bacteriol.*, **152**, 829.
36. Projan, S. J. and Novick, R. P. (1984). *Plasmid*, **12**, 52.
37. Molin, S. and Nordström, K. (1980). *J. Bacteriol.*, **141**, 111.
38. Nordström, K. (1992). *Bio/Technology*, **10**, 661.
39. Cabello, F., Timmis, K. N., and Cohen, S. N. (1976). *Nature*, **259**, 285.
40. Koppes, L. and Nordström, K. (1986). *Cell*, **44**, 117.
41. Tsutsui, H. and Matsubara, K. (1981). *J. Bacteriol.*, **147**, 509.
42. Nielsen, P. F. and Molin, S. (1983). *Plasmid*, **11**, 264.
43. Nordström, K. and Aagaard-Hansen, H. (1984). *Mol. Gen. Genet.*, **197**, 1.
44. Gustafsson, P. and Nordström, K. (1980). *J. Bacteriol.*, **141**, 106.
45. Kessler, D. P. and Rickenberg, H. V. (1963). *Biochem. Biophys. Res. Commun.*, **10**, 482.
46. Light, J. and Molin, S. (1982). *Mol. Gen. Genet.*, **187**, 486.
47. Kupersztoch, Y. M. and Helinski, D. R. (1973). *Biochem. Biophys. Res. Commun.*, **54**, 1451.
48. Cooper, S. and Helmstetter, C. E. (1968). *J. Mol. Biol.*, **3**, 519.
49. Leonard, A. C. and Helmstetter, C. E. (1988). *J. Bacteriol.*, **170**, 1380.
50. Gustafsson, P., Nordström, K., and Perram, J. W. (1978). *Plasmid*, **1**, 187.
51. Nordström, K. (1983). *Plasmid*, **9**, 218.
52. Kornberg, A. and Baker, T. A. (1992). In *DNA replication* (2nd edn), Chapter 18. Freeman & Co., New York.

53. Eliasson, Å., Bernander, R., Dasgupta, S., and Nordström, K. (1992). *Mol. Microbiol.*, **6**, 165.
54. Inuzuki, M. and Helinski, D. R. (1978). *Proc. Natl. Acad. Sci. USA*, **75**, 5381.
55. Itoh, A. and Tomizawa, K.-I. (1980). *Proc. Natl. Acad. Sci. USA*, **77**, 2450.
56. Diaz, R., Nordström, K., and Staudenbauer, W. (1981). *Nature*, **289**, 326.
57. Muraiso, K., Tokino, T., Murotsu, T., and Matsubara, K. (1987). *Mol. Gen. Genet.*, **206**, 519.
58. Clewell, D. B. and Helinski, D. R. (1969). *Proc. Natl. Acad. Sci. USA*, **62**, 1159.

2

Plasmids from Gram-positive bacteria

JUAN C. ALONSO and MANUEL ESPINOSA

1. Introduction

Bacterial plasmids are self-replicating dispensable genetic elements that often specify resistance to antibiotics, heavy metals, bacteriophages, or encode the production of bacteriocins, amino acids, antibiotics, enzymes (e.g. proteases, lipases, catalases, restriction enzymes), etc. Plasmid biology is studied for several reasons. From a basic point of view, plasmids can be used as model systems to study the genetics and biochemistry of cellular processes, and from an applied point of view, they can be used to produce large amounts of industrially important compounds.

In this chapter we describe protocols that can be used to study the transfer and replication of plasmids grouped within the Gram-positive phylum. This phylum appears to consist of four subdivisions, but only two of them are characterized (1). The first one includes species whose chromosomal DNAs contain more than 55% guanine plus cytosine (high G + C content), whereas the second includes species with a low G + C content. The members of the third and fourth subdivisions do not have a Gram-positive cell wall (1).

Little is known about the biology of plasmids from Gram-positive bacteria with a high G + C content, and virtually nothing about those of the third and fourth subdivisions. Recently, circular and linear DNA plasmids from species belonging to the high G + C subdivision have been identified. The linear DNA plasmids, isolated from certain *Streptomyces* species, are unit length, double-stranded molecules, with a protein covalently linked to their 5' end (2). These plasmids replicate by a protein priming mechanism (for an extensive review in this subject see reference 3). The study of plasmids from this high G + C subdivision is beyond the scope of this chapter.

The low G + C subdivision shows at least five major branches, most of which include clostridia. One of the sublines has given rise to four groups of particular interest: *Bacillus, Lactobacillus, Streptococcus* (*Enterococcus, Lactococcus*), the mycoplasma and their clostridial relatives (1). In this chapter we address the plasmid biology of this particular subline.

The protocols presented here, which were mainly developed for plasmids replicating in *B. subtilis*, have been chosen for their versatility. We have assumed that after small modifications these protocols can be adapted for many other Gram-positive bacteria with low G + C content. We are aware, however, that species grouped within this subline are quite different and the adaptation of some of the protocols described here may not be a simple task. For example:

- the mycoplasmas do not have a cell wall and they use the UGA codon for Trp and not as a stop signal
- the species grouped within this subline have quite different physiological requirements

Bacillus species are basically aerobic (although a few also grow anaerobically), *Lactobacillus, Streptococcus*, and the mycoplasmas are essentially anaerobic (but they tolerate and, in some cases, even use oxygen), whereas *Clostridium* species are true anaerobes (1).

Our aim is to describe methods that can be used to establish a plasmid in a given Gram-positive bacterium and to map the genetic information required for plasmid replication and stability in this bacterium. Many of the techniques have been adapted from the published literature. The references cited are either to illustrate the way in which a non-standard technique has been used, or to introduce the reader to a technique not described here.

2. Classification

Plasmids from Gram-positive, low G + C content bacteria have been classified into 11 different families according to the following criteria:

- sequence comparison
- replication mode
- incompatibility grouping

A representative of each family is shown in *Table 1*. Plasmid copy number, which is defined as the number of plasmids per bacterial cell under standard conditions, will not be considered here as a basis for classification because the copy numbers of some plasmids are high in one species, but low in another (4).

The plasmids considered here are double-stranded, covalently closed circular (CCC) DNA molecules, which use either the rolling circle (RC) or theta (Θ) mode for their replication. Linear plasmids have not yet been reported in this group. Four different families have been defined among plasmids which replicate via the RC mode, whereas plasmids using the Θ replication mode fall into seven different families (4–6).

Plasmids replicating via the RC mode usually have a high copy number (at least in the species from which they were originally isolated) and show a

Table 1. Plasmids from Gram-positive low G + C content bacteria

Family	Replication mode	Rep (kDa)	Size (kb)	Phenotype	inc[a] group	Host range	Species
1. pIP501[b]	Θ	57	30.2	Cm,Em,Tra[+]	18	BHR	*Sag*
2. pIP404	Θ	49	10.2	Bc	19	BHR	*Cpr*
3. pTB52[c]	Θ	47	10.8	Tc	17	(NHR?)	*Bsu*
4. pI258	Θ	?	28.2	Pc,Cd,Em	NA	NHR	*Sau*
5. pII147	Θ	?	32.6	Pc,Cd	NA	NHR	*Sau*
6. pCI305[e]	Θ	46	8·7	Cryptic	NA	NHR	*Lla*
7. pAD1[d]	Θ	?	60.0	Hy,Pr,Tra[+]	NA	NHR	*Ehi*
8. pUB110[f]	RC	39	4.5	Nm,Pm	13	BHR	*Sau*
9. pT181[g]	RC	37	4.4	Tc	3	BHR	*Sau*
10. pIM13[h]	RC	17	2.1	Em	12	BHR	*Bsu*
11. pMV158[ij]	RC	24	5.3	Tc	16	VBHR	*Sag*

[a] Plasmid incompatibility was determined in *B. subtilis* as a host. When the plasmid can not be established in this host the *inc* group is denoted as NA (not applicable). The *Sau* plasmids pI524 and pI9789 belong to the same *inc* group as pI258.

[b] pSM19035 (*inc*18), pSM22095 (*inc*18), pAMβ1 (*inc*18), pSM10419, pERL1, pSF9400, and pEG854 are members of the same family.

[c] Orf60 of pEG853 shares homology with RepA of pTB52, hence both are placed within the same family. Plasmids from the pBT52 family replicate in some, but not all *Bacillus* species.

[d] pAMγ1, pBEM10, and pJH2 share the same adhesin (which enable the cell–cell contact), hence they are placed within the same family. The pAD1 adhesin is also immunologically related to those of pPD1, pCF10, pAMγ2, pAMγ3, pOB1, pAM323, pIP1017, pIP1438, pIP1440, pIP1441, pMV120, and pX98.

[e] pSK11L, pSL2, pWV02, pWV03, pWV04, pWV05, and pIL7 belong to the pCI305 family.

[f] pC194 (*inc*8), pAAB1, pAMα1, pRBH1 (*inc*13), pBC16 (*inc*13), pLAB1000, pLP1, pLP3537, pLS11, pST1, pBC1, pUH1, pBS2, pC30il, p353-2, and pVA380-1 belong to the pUB110 family.

[g] pC221 (*inc*4), pUB112 (*inc*9), pS194 (*inc*5), pC223 (*inc*10), pT127 (*inc*3), and pCW7 (*inc*14) belong to the pT181 family.

[h] pSN2, pNE131, pTCS1, pOX1000, pE12 (*inc*12), pE5, and pT48 belong to the pIM13 family.

[i] pE194 (*inc*11), pADB201, pLB4, pFX2, pSH71, pKMK1, and pWV01 belong to the pMV158 family.

[j] Not all members of the pMV158 family present a VBHR.

Abbreviations: RC, rolling circle; Θ, theta; Bc, bacteriocin production; BHR, broad host range; Cd, cadmium ion; Cm, chloramphenicol; Em, erythromycin; Hy, haemolysin production; Nm, neomycin; NHR, narrow host range; Pc, penicillin; Pm, phleomycin; Pr, pheromone response; Tra[+], conjugation proficiency; Tc, tetracycline; VBHR, very broad host range.

Species: *Bsu*, *B. subtilis*; *Bst*, *B. stearothermophilus*; *Cpr*, *Clostridium perfringens*; *Ehi*, *Enterococcus hirae*; *Lla*, *Lactococcus lactis*; *Sau*, *Staphylococcus aureus*; *Sag*, *Streptococcus agalactiae*.

For plasmid references see (4–6, 46).

characteristic modular organization (7). *Figure 1* shows the map of pLS1 (a non-transmissible derivative of pMV158), which is a prototype of plasmids with a very broad host range (VBHR). Not all the members of this family (*Table 1*) have a VBHR. Plasmids replicating via the Θ mode usually have a low copy number and some of them contain partition (*seg*) loci. As a representative of this group, plasmid pDB101 (a derivative of pSM19035) is depicted in *Figure 2*. Unlike other members of the pIP501 family (*Table 1*), pDB101 is non-conjugative and has extraordinarily long inverted repeated sequences that comprise 76% of the molecule (8). Plasmids from Gram-positive bacteria are thought to be randomly segregated (4). However,

methods), with selection for the incoming plasmid (plasmid *a*). Individual transductants are purified and tested for the presence of plasmid *b*. One out of three possible results is obtained from such a test:

(a) Either no transductants are obtained in cells bearing plasmid *b*, or the cells contain only the incoming selected plasmid. In the first case, due to a strong incompatibility, the incoming plasmid can not be established in the presence of a resident plasmid, whereas in the second the entering plasmid can be established and it displaces the non-selected replicon. In both cases they are unable to coexist being therefore incompatible.

(b) Every colony examined contains both plasmids and the copy number of the resident is the same as in the homoplasmid strain. This indicates that both plasmids are compatible and that they do not share common replication functions.

(c) Colonies contain both plasmids, but the resident shows a reduction in its copy number (see *Figure 3*), and will be lost from the cell population after growing the culture for several generations. This result indicates that the donor plasmid exerts a weak incompatibility against the resident one.

The establishment and segregation (incompatibility) tests (*Protocol 1*) are used for such an analysis.

2.2 Incompatibility test

The establishment and segregation (*Protocol 1*) tests are used to measure the ability of a plasmid to displace one that is already resident.

Protocol 1. Incompatibility test

1. Amplify the transducing phage, (e.g. phage SPP1) in wild-type cells bearing the donor plasmid as described in *Protocol 3*.

2. Bacteria (*recA⁻*, desirable but not essential) either bearing the recipient plasmid or plasmid-free, are grown to 1.0×10^8 colonies-forming units (c.f.u.) per ml in rich medium (TY, *Table 2*) supplemented with the appropriate antibiotic (when required). Bacteria and transducing phage are mixed and incubated for 10 min at 37°C (see *Protocol 3*).

3. Wash out the free phages by sedimenting the cells (10 000 *g* for 5 min). Repeat twice and resuspend the cells in fresh medium.

4. Dilute the infected cells (10^{-2} to 10^{-4}) to ensure a suitable number of transductants. Plate samples with selection for the donor or for both plasmids. Incubate the plates overnight at 37°C.

5. Restreak the transductant colonies on plates containing the selective media only for the donor plasmid. After colonies are formed, replica

---

plate them on to plates containing the antibiotic that is selective for the resident plasmid. When high transduction frequencies (10^{-1} to 10^{-3} transductant/survival cells) are expected, the restreaking can be omitted, but in the case of lower transduction frequencies it is necessary to purify the colonies, i.e. to separate the transductants from the amplified phages present in the plate.

6. Test the transductants for the retention of the unselected resident plasmid. When no transductant colonies are obtained, check that under these conditions the plasmid-free cells are transduced by the donor plasmid (establishment test). If weak or no incompatibility is observed, check whether both plasmids are present as independent replicons in the transductants.

7. Transfer individual transductants into liquid media containing the selective antibiotic for donor and/or recipient plasmids and let them grow to early stationary phase.

8. Dilute the culture 10^6-fold in fresh rich medium without antibiotics and grow again to early stationary phase (about 20 generations). Repeat this step once more, dilute the culture to a suitable number of colonies (50 to 100 c.f.u. per plate) and plate the bacteria in the absence of any selective pressure. *Figure 3* shows an example in which selection for the donor plasmid was applied.

9. Streak the colonies on to plates with selection for: both plasmids, the incoming plasmid alone, the resident plasmid, or none of the plasmids.

10. Estimate the percentage of the total colonies carrying the incoming, the resident, and both plasmids. When plasmids appear to be compatible, check for the presence of both plasmids.

11. The test is then repeated by reversing the order of the donor and recipient plasmid.

Two plasmids can be assigned to the same incompatibility group if the donor plasmid is able to displace the resident one or vice versa in the cell population. When incompatible plasmids with different copy numbers are analysed, two types of results are commonly obtained:

(a) When the resident plasmid has a copy number higher than that of the donor plasmid, no transductants are often detected.

(b) If the order is reversed the donor plasmid will be established, but it will displace the non-selected replicon.

When all transductants carry both the donor and the resident plasmid, and both are maintained in the cells independently, we can conclude that both plasmids are compatible. However, when the resident plasmid is lost from

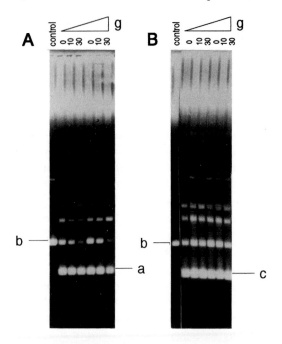

Figure 3. Incompatibility test. A cell culture harbouring plasmid b (resident plasmid) received either plasmids a or c (donor plasmid). Selection was applied and maintained for the incoming replicon only (plasmid a or c). Individual clones were grown to mid-exponential phase (g = 0), diluted into fresh medium, and regrown for a further 10 (g = 10) or 30 (g = 30) generations. Total DNA from these cultures was prepared and analysed for plasmid content in a 1% agarose gel. DNA from the homoplasmid strain containing the resident plasmid was run as control. Plasmid b does not exert incompatibility towards plasmid c, whereas it does against plasmid a (weak incompatibility).

some but not all of the transductants it is advisable to grow the transductants for about 40 generations in the absence of any selective pressure and then streak the resulting colonies on plates with selection for donor, resident, or both plasmids.

When the plasmids to be analysed for their incompatibility grouping share the same marker, it is advisable to introduce a scorable property. This can be done by cloning cartridges (β*gal* or *xyl*E gene) whose products, under certain conditions, give rise to colonies which can be distinguished from the donor one by the generation of a scorable property (e.g. colour).

3. Isolation of plasmid DNA

Isolation of small quantities of plasmid DNAs from bacterial cells is essential for their analysis (see *Protocol 2*). The most reliable and widely used method is based on the alkaline lysis procedure (14). The method can be subdivided into three discrete steps:

(a) The cell wall is partially broken down.

(b) The cells are opened by treatment with SDS/NaOH and the base pairing of linear and open circular DNA is disrupted. High pH (11.5–12.5) causes separation of both strands in the double helix of linear and open circular DNA, but does not do so in the case of supercoiled DNA.

(c) A subsequent neutralization with high salt concentration (0.3 M Na acetate final concentration) at pH 4.5 leads to precipitation of the denatured material (15).

Protocol 2. Isolation of small quantities of plasmid DNA

1. Inoculate from single colonies, in test-tubes containing 3 ml of a rich medium (see *Table 2*) with a selective drug. When low copy number plasmids are to be analysed, inoculate 5 ml of medium. Grow overnight at appropriate temperature.

2. Centrifuge 1.5 ml to 4 ml of the overnight culture in a microcentrifuge (12 000 r.p.m., 1 min). Discard supernatant. When more than 1.5 ml of culture are required use a 2 ml microcentrifuge tube and repeat the centrifugation twice.

3. Resuspend the pellet in 100 μl of solution I (25 mM Tris–HCl pH 8.0, 50 mM glucose, 10 mM EDTA pH 8.0) containing 1 mg/ml of fresh solution of lysozyme (lysostaphin, mutanolysin, or any other cell wall degrading enzyme, or the combination of more than one of them), and 50 μg/ml of RNase A. Mix the solution gently and incubate for 30 min at 37°C. The stock solution of RNase A (at a concentration of 10 mg/ml in 5 mM Tris–HCl pH 8.0) is made DNase-free by incubation at 100°C for 15 min.

4. Add 200 μl of solution II[a] (0.2 M NaOH, 1% (w/v) SDS). Mix the suspension by inverting the tube several times until the culture clears. Keep the tube on ice for 5 min.

5. Add 150 μl of solution III (3 M Na acetate, pH 4.5). Mix the suspension by inverting the tube several times until the gross precipitate separates. Keep the tube on ice for 15 min. If the pH is not correct, precipitates will not form or will do so poorly. Sodium acetate is prepared by adding glacial acetic acid to 3 M Na acetate until pH 4.8 is obtained.

6. Centrifuge in a microcentrifuge at 12 000 r.p.m. for 5 min. Pour the clear supernatant into a fresh microcentrifuge tube.

7. Add to the supernatant of step **6**, 1 ml of cold 96% ethanol (at −20°C). Mix by inverting the tubes several times and keep at −20°C for 10 min.

Protocol 2. *Continued*

8. Pellet the precipitated DNA by centrifugation in a microcentrifuge at 12 000 r.p.m. for 4 min. Discard the supernatant.

9. Dissolve the pellet in 180 μl water and 20 μl of solution III. Reprecipitate with two volumes of cold ethanol. Keep for 10 min at −20°C.

10. Pellet the DNA as in step **8**.

11. Dry the pellet under vacuum (desiccator or vacuum centrifuge) and resuspend in 30 μl water. Between 0.5 μg to 5 μg of plasmid DNA are usually obtained from 1.5 ml (high copy number plasmids) or 5 ml (low copy plasmids) cell cultures.

12. Pour distilled water or TE buffer into a Petri dish (about 10 ml). Take a membrane filter (e.g. Millipore type VS, 0.025 μm) with flat forceps and mark it with a waterproof pencil at the position of 12 o'clock. Place it on the liquid with the mark upwards. Place the DNA of step **11** on the filter and keep dialysing for 10 min. Up to five samples can be dialysed simultaneously. Collect the DNA and store at 4°C.

[a] Note that when plasmids to be purified have a G + C content below 35% it is convenient to reduce the NaOH concentration to 0.17 M.

Unlike the high copy number plasmids from *E. coli*, the plasmids described here require an unstable plasmid-encoded product for initiation of plasmid DNA replication. Therefore, they can not be amplified by addition of protein synthesis inhibitors, such as chloramphenicol.

The protocol yields plasmid DNA of sufficient quality for use in most enzymatic manipulations. When large quantities are required, the alkaline lysis method can be scaled-up and it is followed by CsCl/ethidium bromide (EtBr) centrifugation as described elsewhere (15–17).

4. Host range and plasmid transfer

Bacteriophages and plasmids are the major vehicles for the transfer of genetic information between bacteria. The natural occurrence of these transfers plays an important role in the evolution of bacterial populations.

We present here some systems used by certain NHR, broad host range (BHR), and VBHR plasmids. We consider that a given plasmid has a BHR when it is able to replicate in more than two different species of the same subline, and VBHR when it is also able to replicate in bacteria classified in a different phyla (e.g. purple bacteria). A question that remains unanswered is what is/are the basis(es) for the BHR of a plasmid. There does not appear to be a simple answer to this question (6, 18).

Natural (transformation of competent cells, phage-mediated plasmid transduction, or conjugation) or artificial methods (electroporation or protoplast transformation) can be used to introduce plasmids into Gram-positive bacteria. Among the natural methods, conjugation offers the widest spectrum of transfer. Recently, Trieu-Cuot *et al.* (19) have documented the conjugative transfer from *E. coli* to different species of Gram-positive bacteria with low G + C content. This observation provides an insight into the considerable potential for gene distribution throughout the bacterial kingdom. The method of choice will depend on the species and on the specific purpose of the experiment.

4.1 Transformation of competent cells

Competence is a metabolic state in which a cell develops the ability to take up naked DNA from the environment and to transport it through the cell wall–membrane complex as a single-stranded (ss) DNA. Several growth media has been developed to induce competence. Transformation of competent cells has been studied intensively in *Streptococcus pneumoniae*, *S. gordonii* (formerly *S. sanguis*), *S. oralis*, and *B. subtilis*. Recent reviews dealing with the modes of gene transfer have been published (20). In short, competent cells can bind, very efficiently, several molecules of double-stranded (ds) DNA, in a manner that is independent of nucleotide sequence. The dsDNA bound by the cells is cleaved by nucleases into fragments with an average size of 15 to 30 kb (20) and transported inside the cell as ssDNA molecules. Entry of a monomeric plasmid molecule would give rise at most to a linear ssDNA, which could not be recircularized. For plasmid establishment, recombination between donor ssDNA molecules seems to be needed. This allows recircularization and replication of the plasmid genome. Thus, plasmid monomers are a very poor substrate for transformation if at all, whereas oligomers can readily transform competent cells (21). Furthermore, phosphatase-treated plasmid DNA is unable to transform competent cells. The plasmid transformation activity can be increased 50 to 100-fold if homology between the donor and recipient DNA is provided.

Due to the large physiological differences between species that develop natural competence a common protocol is not available. Detailed protocols, described by Anagnostopoulos and Spizizen (22); Lacks (23); Rottländer and Trautner (24); Pakula and Walczak (25); Ronda *et al.* (26) are still in use.

4.2 Transduction

Transfer of genetic information among bacteria by bacteriophage-mediated transduction is another relevant mode of natural plasmid transfer. Plasmids, which are usually smaller in size than a phage genome, can be encapsidated into phage proheads to produce plasmid-transducing particles. The encapsidation of linear plasmid concatemers has been detected in all bacteriophage systems tested so far.

Plasmid transduction frequencies are generally low, about 10^{-5} to 10^{-7} per surviving cell. However, when homology (as little as 50 bp) is provided between the phage genome and the plasmid, the transduction frequency increases up to 10^5-fold (transduction facilitation effect). The mechanism leading to the production of packageable plasmid DNA concatemers after phage infection has not yet been totally elucidated. Experimental evidence obtained from *E. coli* and *B. subtilis*, however, points to the involvement of a recombination-dependent concatemeric plasmid replication mechanism with a subsequent formation of a plasmid:phage chimera. The chimeric particle is then recognized by the phage packaging machinery and encapsidated into phage proheads (27).

Transduction can be used to transfer plasmids from easily transformable *B. subtilis* 168 to strains which are refractory to transformation or poorly electro-porated. The limitation of the system is the range of sensitivity of the bacteria to a given bacteriophage. The procedure described in *Protocol 3* can be used to mobilize plasmids among different bacteria. Our procedure uses bacilli as a host, but the same principle can be used for phages of different genera. The phages more commonly used are: SPP1 (*B. subtilis* 168, *B. natto*, and *B. niger*), SPO2 (*B. subtilis* 168, and *B. amyloliquefaciens*), CP-51 (*B. anthracis*, *B. cereus*, and *B. thuringiensis*) and β22 (*B. subtilis* 168, *B. subtilis* W23, *B. niger*, *B. natto*, *B. pumilus*). Here we describe a protocol to generate SPP1 transducing particles, but the same procedure can be used for many other lytic phages. When a lysogenic phage is to be used to mobilize a plasmid, it may require prophage induction.

Protocol 3. SPP1 mediated plasmid transduction

1. Prepare ten-fold serial dilutions of a SPP1 stock in TY (*Table 2*). Dispense 0.1 ml of each dilution into a sterile 5 ml tube (to ensure a suitable number of plaque-forming units (p.f.u.) per plate). Add 0.1 ml of exponentially growing *B. subtilis* cells, mix by vortexing, and incubate for 5 min at 37°C to allow the bacteria to absorb the phages. Add 3 ml of molten agar (TY soft agar), mix, and pour the entire contents on to the centre of a pre-warmed TY agar plate. Let it stand for about 5 min at room temperature to allow the soft agar to solidify, invert the plate, and incubate at 37°C. Plaques should be counted after 12 to 18 h. Stocks prepared from liquid cultures contain about 1.0×10^{10} to 5.0×10^{10} p.f.u./ml.

2. Grow *B. subtilis* cells bearing plasmid (donor plasmid) at 37°C to an OD_{560} of 0.7 (about 1.0×10^8 c.f.u./ml) in 5 ml of TY medium supplemented with the appropriate antibiotic, and infect the cells with three phages/bacterium.

3. Add $CaCl_2$ to 50 mM final concentration (a white precipitate is formed) 5 min after infection.

4. Incubate the infected culture for 3 h with agitation at 37°C.

5. Remove cell debris by centrifugation at 10 000 *g*, 10 min, 4°C. Recover the supernatant and store at 4°C. The titre of the stock should be about 10^{10} p.f.u./ml and should remain stable for several years.

6. Grow 5 ml of *B. subtilis* cells at 37°C to an OD_{560} of 0.7 (about 1.0×10^8 c.f.u./ml) in TY medium, infect the bacteria at multiplicity of infection of three, and incubate for 20 min at 37°C with agitation.

7. Pellet the infected cells to remove free phages. Repeat this step twice. It is recommended to use a phage antiserum to remove the free phages if available.

8. Prepare a ten-fold serial dilution of the infected cells and plate 0.1 ml from each dilution to ensure a suitable number of transductants being present in one of the selective plates. The titre of infected and surviving cells should be also determined. Transduction is usually expressed as the number of colonies resistant to the donor-plasmid marker (transductants) per surviving cells or infecting particles.

4.3 Conjugation

Conjugative plasmids are able to mediate their own transfer, or the transfer of a mobilizable plasmid, by cell-to-cell contact. Conjugative plasmids have been found in various Gram-positive species and they fall into two classes. One class, which includes the NHR plasmids of the pAD1 family (see *Table 1*) or certain lactose plasmids from *Lactococcus lactis*, transfers at high frequency ($\sim$ 1% of donors) to their recipient cells. The sex pheromones secreted by plasmid-free recipient cells of *Enterococcus hirae* (formerly termed *Streptococcus* or *Enterococcus faecalis*) are responsible for such high transmission (the lactococcal conjugation system has many features in common with the *E. hirae* transfer system, but little is known about this system). As a response to sex pheromones, the *E. hirae* donor strain synthesizes an adhesin which allows the cell-to-cell contact needed for conjugative transfer (28, 29). So far, this interesting class of transfer system is confined to NHR plasmids. The other class, which includes plasmids of the pIP501 family, transfers poorly in broth but allows mating when cells are concentrated on filter membrane (0.1% of donors). Plasmids of this class can be conjugatively transferred and they also mobilize other plasmids among a broad range of Gram-positive species with low G + C content. This class of conjugative plasmids provides the method of choice when interspecies plasmid transfer is of interest. A procedure to transfer plasmids among different Gram-positive bacteria is presented in *Protocol 4*.

Protocol 4. Conjugative plasmid transfer

1. Grow donor strains carrying pIP501 (Em^r, Cm^r, see *Table 1*) overnight in rich medium containing the appropriate selection antibiotic. The recipient strain should grow under the same condition with selection for the host markers.
2. Pellet the donor strain and resuspend the cells in fresh medium lacking antibiotic.
3. Mix in a ratio of one donor to two recipient bacteria, filter through a 0.45 μm pore size membrane filter, and wash with 1 ml of 1 × S-Base (see *Table 2*).
4. Place the filter, bacteria uppermost, on rich agar medium and incubate for 18 h at 37°C.
5. Transfer the filter to a test-tube and resuspend the bacteria in a fresh rich medium by vortexing.
6. Serially dilute the cell suspension, plate from each tube of the dilution series to ensure a suitable number of transconjugants on to the appropriate solid media, and incubate at 37°C. Each mating experiment includes sets of controls that select for either donor or recipient strain.

Certain strains of streptococci contain sets of genes capable of mediating their own conjugative transmission that are not located on plasmids but rather on their chromosomes. These elements, termed conjugative transposons, have a VBHR, since they are capable of self-transfer as well as the mobilization of plasmids in Gram-negative and Gram-positive bacteria (30). The first step in this kind of mobilization is the excision and circularization of the transposon, which is then followed by a plasmid-like conjugational transfer (31).

4.4 Electroporation

Electroporation is a method in which a high voltage electric pulse passes through an aqueous suspension of DNA mixed with bacteria. The method is particularly useful for introducing DNA into a bacterium which is either refractory to competence or poorly competent. Many different electroporation methods have been published to transform bacilli, lactococci, enterococci, and streptococci (32). The most important parameters appear to be the growth media (especially the presence of glycine to inhibit cell wall synthesis), growth phase, and electroporation conditions (field strength, DNA concentration, electroporation solutions). In *Protocol 5*, we present a procedure that can be used for the different species grouped within the Gram-positive bacteria with low G + C content. However, the procedure can be used as a starting point to develop a method for other organisms of interest.

Protocol 5. Electroporation

1. Grow cells overnight in Yec-M9 plus glycine[a] (at concentrations ranging from 0.4% to 10%), TBAB, or GM17[b] (plus 40 mM DL-threonine) (see *Table 2*).

2. Dilute the overnight culture into fresh medium (in the case of Yec-M9 the medium should also contain glycine) to bring the OD_{560} to 0.05, and incubate the cells for about 60 min to 120 min at 37°C.

3. Chill the culture in iced-water (10 min) and harvest the cells by centrifugation (10 000 g, 5 min, at 4°C). Discard the supernatant.

4. Wash the cells three times in half the volume of the chilled SMG solution (500 mM sucrose, 1 mM $MgCl_2$, adjusted to pH 5.0 with HCl). Pellet the cells after each washing step by centrifugation (10 000 g, 5 min, 4°C), and resupend in 1/25 to 1/100 of the original volume of the SMG solution.

5. Incubate the cells on ice for 60 min and then dispense in 0.5 ml aliquots in microcentrifuge tubes. Submerge the tubes in dry-ice–ethanol and store in a −70°C freezer. The cells remain competent for at least a year.

6. Thaw the frozen cells in an ice–water-bath just prior to use.

7. Add cells from step **6** to the electroporation cuvette (800 μl for 0.4 cm (inter-electrode gap) and 50 to 100 μl for 0.2 cm cuvettes). Add 1 μl to 10 μl of DNA containing 0.01 μg to 1 μg of DNA in water and mix well by vortexing.

8. Electroporation is carried out using the 25 μF setting on the Bio-Rad Gene Pulser. When the 0.2 cm cuvette and the pulser controller units are used, the resistance should be set at 200 Ω and the field strength should be 7.5 to 12.5 kV/cm. Take the cells out of the cuvette immediately after electroporation.

9. Dilute the cells 1/2 to 1/10 on GM17 (plus 30 mM sucrose) or TBAB medium (containing a sub-inhibitory concentration of the antibiotic if the resistance gene is inducible) and plate on GM17 (plus 30 mM sucrose) or TBAB plates containing the appropriate antibiotic for the selection of transformants.

[a] Note that for certain species of bacilli and streptococci the presence of glycine is detrimental.
[b] GM17 is recommended for lactococci.

4.5 Transformation of protoplasts

Protoplasts are prepared by cell wall removal in media supplemented with osmotic stabilizers such as sucrose, succinate, mannitol, or sorbitol. A variety of methods exist for protoplast transformation, but they mainly differ in the enzymes for cell wall degradation and/or the media used for cell wall regeneration. In our hands, sucrose-based regeneration media give the highest reproducibility and lowest regeneration time (overnight). The protocol presented here (*Protocol 6*), originally developed by Chang and Cohen (33), was adapted by Puyet *et al.* (34) for *B. subtilis* cells.

When establishment of a plasmid in a given Gram-positive bacterium other than *B. subtilis* is desired we recommend the use of combinations of enzymes for cell wall removal (e.g. lysozyme, lysostaphin, mutanolysin, any other equivalent). An efficiency of 5% to 10% of regenerated cells is usually obtained. Note that the regenerated colonies may show a viscous appearance due to secretion of saccharolytic enzymes.

Protocol 6. Protoplast transformation[a]

1. Grow 5 ml of culture overnight in rich medium (e.g. penassay broth (PAB), antibiotic medium No. 3, Difco).

2. Dilute (1:50) the overnight culture into 50 ml of fresh pre-warmed medium, and incubate at 37 °C until about 1.0×10^8 c.f.u./ml.

3. Harvest the cells by centrifugation (10 000 *g*, 5 min, at 4 °C), and resuspend in 5 ml of SMMP (prepared by mixing equal volumes of 4 × PAB with 2 × SMM, see *Table 2*).

4. Add lysozyme (2 mg/ml) and mutanolysin (20 U/ml), and incubate while shaking gently for 45 to 60 min at 37 °C. Check for protoplast formation by phase-contrast microscopy.

5. Centrifuge the protoplast suspension (2000 *g*). Wash the pellet with 5 ml of SMMP and resuspend in the same volume of SMMP. The protoplasts can be stored at −80 °C without the addition of glycerol.

6. Mix 200 μl of protoplast suspension with the desired plasmid DNA (0.5 μg to 1 μg, well dialysed against water, see *Protocol 2*, step **12**) and immediately add 400 μl of PEG 40% (*Table 2*) to the mixture. Mix the contents of the tube and incubate for 3 min at 4 °C.

7. Add 2 ml of SMMP to dilute the PEG mixture and harvest the protoplasts by centrifugation (2000 *g* for 10 min). Discard the supernatant.

8. Resuspend protoplasts into 1 ml SMMP and incubate for 90 min at 37 °C to allow phenotypic expression of the antibiotic resistance marker.

9. Make serial dilutions into SMMP and plate on to vy-R5 plates (containing the appropriate antibiotic) for regeneration. Plate 1 μl and 10 μl of the original protoplast suspension in TBAB plates to estimate the number of non-protoplasted cells, referred to as the c.f.u. of the original culture.

10. Incubate for 16 h to 20 h at 37 °C.

[a] Use detergent-free glassware throughout.

---

5. Plasmid replication modes

Two modes of replication, which are thought to be unlinked to recombination, have been reported for the plasmids from the Gram-positive bacteria under characterization. In addition, a recombination-dependent plasmid replication mechanism has been described (see below). In the theta mode, the sites for priming leading and lagging strand syntheses are located close to each other within the origin of replication. From this origin, replication proceeds either uni- or bidirectionally and terminates at a sequence-specific replication terminus at or near the replication origin. The relaxed concatemeric dimer is then resolved into monomeric rings (35). Theta replication is used by the first seven families of plasmids shown in *Table 1*. The first two families have a BHR, whereas the remaining five have a NHR. Plasmids of the pIP501, pIP404, pTB52, and pCI305 family code for an essential *trans*-acting replication protein (Rep). The biochemical activity of the Rep protein is unknown.

The RC replication cycle can be divided into two stages which mechanistically, resemble the DNA replication of single-stranded *E. coli* phages (36). In the first stage, the replication of the leading strand is accomplished requiring, besides host functions, a plasmid-encoded initiator–terminator protein (Rep). In the second stage, the complementary strand (lagging strand) is synthesized. This may be initiated as soon as the plasmid single-stranded origin (sso) is exposed in its ssDNA form or after synthesis of the leading strand has been completed.

Plasmid leading strand synthesis can be subdivided into three steps (initiation, elongation, and termination).

(a) *Trans*-acting and rate limiting Rep protein binds at the plasmid double-stranded origin (dso) region. No high energy co-factor is required but the substrate must be negatively supercoiled. The Rep protein makes a single-strand DNA cleavage at dso and might remain attached to the 5' terminus during strand displacement. The Rep protein may provide a 'functional gap' which allows a 3' to 5' DNA helicase to bind and initiate unwinding of DNA.

(b) The 3' hydroxyl (primer) end at the nicking site is then extended by DNA polymerase III.

(c) After one full round of replication, Rep terminates the strand displacement by cleaving the regenerated dso to produce unit length single-stranded circular [SS(c)] DNA similar to the φX174 bacteriophage (see 35).

Those plasmids grouped within families eight to eleven use this replication mode (4–6).

Recently, a recombination-dependent DNA replication mode was also detected for the plasmids considered here. This mode, which appears to share some common properties with late phage replication, leads to the accumulation of linear concatemeric (LC) plasmid molecules (termed also hmw DNA). The genetic requirements, as well as a model for plasmid concatemeric replication have been recently reviewed (27). The generation of concatemeric plasmid DNA by this replication mode will therefore not be discussed further here.

5.1 Analysis of the plasmid replication mode

The analysis of plasmid replication intermediates (RIs) is a direct way to determine the plasmid replication mode, the origin of replication, and whether replication is uni- or bidirectional.

Plasmid RC replication generates RIs that are unit length single-stranded DNA circles [SS(c)], which corresponds to only one of the two DNA strands (the leading strand) (37–39). Cells bearing plasmids that have been increased in size (e.g. carrying DNA inserts) or which have very efficient sso utilization accumulate a very small amount of SS(c) DNA. In many cases, lagging strand synthesis is mediated by RNA polymerase. For this reason, the addition of 10 µg/ml of rifampicin to growing cells, ten minutes before DNA purification, markedly increases (three to five-fold) the level of RIs, simplifying the detection of SS(c) DNA (40). The plasmid forms are then separated by agarose gel electrophoresis and the SS(c) DNA is detected by hybridization.

In the case of theta-replicating plasmids, the RIs can be separated from non-replicating molecules either by sucrose sedimentation gradient, dye–buoyant density centrifugation (see *Protocol 7*), two-dimensional (2D) gel electrophoresis (see *Protocol 8*), or by the combination of these techniques.

5.1.1 Identification of rolling circle replication

The presence of SS(c) DNA is an important criterion for identifying RC replication (37), but its detection is not always easy. For example, the accumulation of SS(c) DNA is inversely proportional to the efficiency with which lagging strand DNA synthesis initiates. Lagging strand DNA synthesis may be initiated after the leading strand synthesis is completed (low efficiency of initiation) or as soon as the lagging strand replication origin is in a single-stranded form (high efficiency).

The procedure described here to identify SS(c) DNA is the one used by Riele *et al.* (37) with minor modifications. In short:

(a) Add rifampicin (10 µg/ml) to a plasmid-bearing culture growing at mid-exponential phase, and allow to grow for a further 10 min. Collect and lyse the cells.

(b) The different plasmid DNA forms from a whole cell lysate are separated by agarose gel electrophoresis (usually two gels are run in parallel).

(c) DNA in one of the gels is alkali denatured prior to its transfer to a nylon membrane, whereas the DNA in the second gel is blotted without denaturation (for detection of ssDNA).

(d) The blotted DNA is hybridized with a radiolabelled strand-specific probe. The ssDNA nature of IRs can be confirmed by treating the DNA preparation with endonuclease S1 prior to DNA electrophoresis.

The same type of protocol can be used for the detection of linear concatemeric (LC) plasmid molecules. Such plasmid DNA molecules accumulate in certain genetic backgrounds and/or in wild-type cells bearing genetically engineered plasmids which replicate by the RC mode (27).

5.1.2 Isolation of theta replication intermediates

The buoyant density of the DNA is dependent on its G + C content, but the binding of an intercalating dye, such as EtBr or propidium diiodide (PD) reduces its buoyant density. Linear or nicked circular DNA binds more PD or EtBr than CCC DNA. During replication via theta intermediates, the DNA remains covalently closed, but the regions in which both strands are being synthesized are partially relaxed. Replication intermediates will therefore bind an amount of PD which is intermediate between that for non-replicating CCC DNA and that for an open circular molecule, allowing these RIs to be isolated on a PD/CsCl dye–buoyant density gradient. After isopycnic centrifugation, the CsCl gradient has the following appearance. Two major bands of DNA are located in the centre of the tube. The upper band contains the chromosomal DNA and relaxed plasmid forms, whereas the lower band contains the CCC plasmid DNA. The material above the chromosomal DNA band are proteins and the deep red pellet at the bottom of the tube consists of RNA complexes. The RIs DNA band is between the chromosomal and the CCC plasmid DNA bands.

Protocol 7. Isolation of replication intermediates

1. Grow 2 litre culture of bacteria bearing the plasmid of interest in minimal medium to late exponential phase at the lowest temperature which is compatible with active cell growth and plasmid yield (between 30 °C and 35 °C). This is to reduce the rate of replication fork movement thus increasing the replication time.

2. Pour the cells on to an equal volume of crushed-ice minimal medium containing a sodium azide solution (10 mM final concentration) to stop cell growth as rapidly as possible. When the cell suspension is cold, pellet the bacteria (6000 g, for 10 min) and resuspend them in 20 ml (1/100 volume) of a suitable buffer (50 mM Tris–HCl pH 8.0, 50 mM sucrose, 10 mM EDTA, buffer A). Add 2 ml of a 2 mg/ml fresh solution of lysozyme and 1 mg/ml of RNase A in buffer A. RNase A, at a concentration of 10 mg/ml in 5 mM Tris–HCl pH 8.0, is made free of DNases by incubation at 100 °C for 15 min. Mix the solution gently and incubate it for 30 min at 37 °C. Add 2.5 ml of 0.5 M EDTA pH 8.0. Seal

---

of segregational instability in the absence of selective pressure (3–8% per generation for the low copy variant; two orders of magnitude less unstable for the high copy variant), which is not affected when the plasmids carry large inserts. The stability of the recombinants may be related to the θ mode of replication of pAMβ1. Recombinant plasmids carrying the nisin resistance determinant were obtained by direct cloning in *L. lactis* (4). A disadvantage of these vectors is that the presence of insert DNA can not be judged on the basis of a simple test such as marker inactivation.

ii. pIP501-derived vectors

In a number of instances the vector pGB301 has proven to be useful in transformation of lactococci. This vector arose as a spontaneous deletion mutant from the *Streptococcus faecalis* strain JH2-2 plasmid pIP501, that carries an inducible Em^r gene and a Cm^r determinant, pGB301 (9.8 kb), retained these antibiotic resistance markers (5). Although the vector accepted large DNA inserts in the *Hpa*II site, marker inactivation (Cm^r) was not observed.

iii. Dual replicon vectors

Before the development of gene cloning vectors constructed from cryptic lactococcal plasmids that proved to be promiscuous with respect to host specificity (see Section 2.1.2), dual replicon shuttle vectors were constructed, that have been used to some extent in lactococci. These vectors include pVA838 (9.2 kb) (6) and pSA3 (10.2 kb) (7), (see *Table 1*). Other vectors suitable for use in lactococci are the pMU1327 and pMU1328 plasmid pair (these two vectors differ only in the orientation of the MCS) which can be exploited as promoter screening vectors (8).

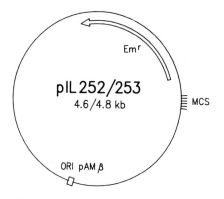

Figure 1. The general cloning vectors pIL252 and pIL253. Unique restriction enzyme recognition sites in the multiple cloning site (MCS): *Eco*RI, *Sma*I, *Bam*HI, *Sal*I, *Pst*I, *Sac*I, *Xba*I, *Xho*I.

Table 1. Heterologous cloning vectors

Vector	Replicon	Size (kb)	Antibiotic resistance gene(s)	MCS	Marker inactivation	Reference
pIL252	pAMβ1	4.7	Em	Yes	No	4
pIL253[a]	pAMβ1	5.0	Em	Yes	No	4
pGB301	pIP501	9.8	Em Cm	No	No	5
pVA838[b]	pACYC184 pVA749	9.2	Em Cm	Yes	Yes	6
pSA3[c]	pACYC184 pGB301	10.2	Em Cm Tc	No	Yes	7
pMU1327/1328[d]	pVA838	7.5	Em	Yes	No	8

[a] High copy number plasmid in *L. lactis*. All other vectors in the table are of the low copy number-type.
[b] Cm^r gene inactivated by insertion in the unique *Eco*RI and *Pvu*II sites.
[c] Cm^r gene inactivated by insertion in the unique *Eco*RI and *Eco*RV sites; Tc^r gene inactivated by insertion in the unique *Bam*HI, *Sal*I, and *Sph*I sites. Recombinant plasmids are selected in *E. coli* and transferred from this host to lactococci.
[d] Promoter screening vector, contains promoterless *cat-194* and lambda t$_0$ terminator.

2.1.2 Cloning vectors derived from lactococcal replicons

After the demonstration that the smallest plasmid (pWV01) of the plasmid complement of *L. lactis* subsp. *cremoris* strain Wg2 was capable of replication in *Bacillus subtilis*, this cryptic plasmid was provided with the chloramphenicol acetyl transferase gene of pC194 and the erythromycin resistance gene of pE194, giving rise to the basic cloning vector pGK12 (9) (*Figure 2*). pGK12 proved to be a promiscuous vector, also capable of replicating in *E. coli*, thus providing a vehicle in which recombinants can be obtained in lactococci, *E. coli*, and *B. subtilis*. The use of the two latter hosts has facilitated the genetic analysis of lactococci, because use could be made of the advanced recombinant DNA protocols available for these hosts. A second advantage of this plasmid is that the presence of inserts can be readily detected by cloning in the unique restriction sites available both in the Cm^r and Em^r gene of the vector. In contrast to the pAMβ1-derived vectors, those derived from pWV01 replicate according to the rolling circle mode. Recombinant vectors carrying inserts in excess of 10 kb tend to be segregationally unstable (10).

Plasmid pWV01 has been sequenced and proved to be almost identical to the cryptic plasmid pSH71 that was genetically marked in a way similar to pWV01 to produce pNZ12 (11) and pCK1 (12). To improve the versatility of this type of vector, both pNZ12 and pGK12 were equipped with a multiple cloning site (MCS), giving rise to pNZ123 (G. Simons, personal communication) and pGKV21 (13), respectively (*Figure 3*). In contrast to pNZ123 (50 copies per chromosome equivalent), pGKV21 is a low copy number vector in

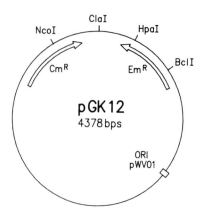

Figure 2. Broad host range cloning vector pGK12.

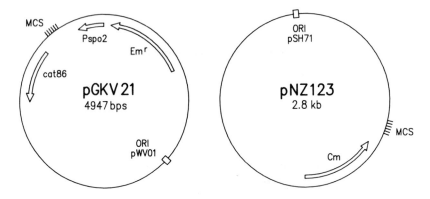

Figure 3. The cloning vectors pGKV21 and pNZ123. Unique cloning sites in MCS of pGKV21: *Bam*HI and *Sal*I. Unique cloning sites in MCS pNZ123: *Sau*3A, *Sca*I, *Eco*RI, *Xba*I, *Sac*I, *Hind*III, *Xho*I.

lactococci (10 copies per chromosome equivalent), but has a high copy in *E. coli* (100 copies per chromosome equivalent). Several other vectors have been constructed including some which are based on pWV01-like cryptic lactococcal plasmids (see *Table 2*).

The analysis of lactococcal transformants obtained with these (recombinant) plasmids is greatly facilitated by the use of a small scale plasmid DNA isolation procedure, adapted from the 'miniprep' method of Birnboim and Doly (14) as described for *E. coli*. The procedure for *L. lactis* is described in *Protocol 2* and yields DNA that can be efficiently digested with restriction enzymes.

Protocol 2. 'Miniprep' plasmid DNA isolation procedure by means of the alkaline lysis method

Materials

- GM17: M17 (37.25 g/litre; Difco, Detroit, USA), 0.5% (w/v) glucose
- 1 M DL-threonine
- demineralized water
- solution A: 20 mM glucose, 10 mM Tris pH 8.0, 10 mM EDTA, 50 mM NaCl
- solution B: 1% (w/v) SDS, 0.2 M NaOH
- lysozyme
- mutanolysin (1500 U/ml; Sigma, St. Louis, USA)
- 3 M sodium acetate pH 5.2
- phenol saturated with TE
- CHL/IAA: chloroform/isoamylalcohol (24:1)
- 96% ethanol (−20°C)
- 70% ethanol (−20°C)
- TE: 10 mM Tris pH 8.0, 1 mM EDTA pH 8.0
- RNase in TE (10 mg/ml)
- sterile pipette tips and 2 ml plastic tubes (Eppendorf)
- microcentrifuge
- vacuum centrifuge or desiccator

Method

1. Inoculate the strain to be examined in 3 ml GM17 containing 40 mM DL-threonine and selective antibiotics. Grow overnight at 30°C.

2. Pellet 2 ml of the culture in a 2 ml tube by centrifugation (10 000 r.p.m. 3 min).

3. Resuspend the pellet in 1 ml of demineralized water and pellet the cells by centrifugation (10 000 r.p.m., 3 min).

4. Resuspend the pellet in 200 μl solution A + lysozyme (5 mg/ml).

5. Add 5 μl mutanolysin.

6. Incubate 30 min at 37°C.

7. Add 400 μl solution B. Mix by inversion or mild vortexing. Solution should become clear. Store at room temperature maximally 5 min.

8. Add 300 μl 3 M sodium acetate pH 5.2. Mix by inversion or mild vortexing. Store at room temperature or on ice for minimally 5 min.

9. Pellet the precipitate by centrifugation (10 000 r.p.m., 10 min).

10. Transfer the supernatant to a clean 2 ml tube (discard the tube with the pellet).

11. Add 400 μl phenol to the supernatant and vortex for at least 10 sec.

12. Add 400 μl CHL/IAA and vortex for 10 sec.

13. Separate the two phases by centrifugation (10 000 r.p.m., 5 min).

14. Transfer 650 μl of the aqueous (upper) phase to a clean 2 ml tube.

15. Precipitate DNA by adding 1300 μl 96% ethanol, leave at room temperature for 10 min.

Protocol 2. *Continued*

16. Pellet the DNA by centrifugation (10 000 r.p.m., 10 min) and remove supernatant.

17. Rinse the pellet gently with 1 ml 70% ethanol, centrifuge (10 000 r.p.m., 2 min), and remove supernatant.

18. Dry the pellet under vacuum (desiccator or vacuum centrifuge) and dissolve in 25 μl TE.

19. Add 1 μl of RNase.

20. Use 5 μl samples for high, or 10–15 μl samples for low copy number plasmids for agarose gel electrophoresis.

21. Store the DNA at 4°C or −20°C.

If the correct clone has been identified, plasmid DNA can be purified from *L. lactis* on a large scale by simply scaling up the volumes given in *Protocol 2*. After step 10 the DNA can be precipitated with alcohol, followed by a standard CsCl/EtBr centrifugation to purify the plasmid DNA (15).

2.2 Special purpose vectors

Several special purpose vectors have been constructed for lactococci, including those which allow screening for transcriptional signals, (inducible) expression of genes, secretion of proteins, and (food-grade) integration of genes (see *Table 3*). A number of these vectors, most of which are based on pWV01 or pWV01-like replicons, are discussed here in more detail.

2.2.1 Transcription signal-screening vectors

The majority of the vectors constructed that allow the selection of DNA fragments carrying promoters active in lactococci can be used for the assessment of promoter strengths in a variety of hosts, because of the broad host range of their replicons. By slightly modifying these promoter screening vectors, they can also be used for the isolation of transcriptional terminators, or to assess the functionality of suspected terminators.

i. Promoter screening vectors

In all promoter screening vectors a probe gene is incorporated from which the native promoter has been deleted, but in which the ribosome-binding site (RBS) is still intact. A suitable restriction site (or a MCS) is added for the insertion of DNA fragments to be probed. In most promoter screening vectors the probe gene is a gene potentially conferring resistance to chloramphenicol, either obtained from *Bacillus pumulis* (*cat-86*) or from *Staphylococcus aureus* (*cat-194*). Because of the promiscuity of the pWV01 replicon the presence and functionality of promoters can be screened in a wide variety of hosts, including both Gram-positive and Gram-negative bacteria. The level

Table 2. General cloning vectors based on small cryptic lactococcal replicons

Vector	Replicon	Size (kb)	Markers	MCS	Remarks	Reference
pGK12	pWVO1	4.4	Em Cm	−	Marker inactivation: *Nco*I site in Cm^r gene; *Hpa*I and *Bcl*I in Em^r gene.	9
pGK13	pWVO1	5.0	Em Cm	−	As pGK12; additional unique cloning sites *Hind*III, *Eco*RV, *Nhe*I, *Bam*HI, *Eco*RI.	(J. Kok, personal communication)
pGKV21	pWVO1	4.9	Em Cm	+	Marker inactivation: *Bcl*I site in Em^r gene.	13
pNZ12	pSH71	4.3	Cm Km	−	Marker inactivation: *Nco*I site in Cm^r; *Bgl*II in Km^r	11
pNZ17^a	pSH71	5.7	Cm Km	+	Marker inactivation: *Nco*I site in Cm^r; *Bgl*II in Km^r	16
pNZ18/19	pSH71	5.7	Cm Km	+	Marker inactivation: *Nco*I site in Cm^r; *Bgl*II in Km^r.	(W. M. De Vos, personal communication)
pNZ123	pSH71	2.8	Cm	+	No marker inactivation.	(G. Simons, personal communication)
pCK1	pSH71	5.5	Cm Km	−	Marker inactivation: *Bgl*II in Km^r. Additional unique cloning sites: *Ava*I, *Bam*HI, *Eco*RI, *Pvu*II, *Xba*I.	12
pVS2	pSH71	5.0	Em Cm	−	Marker inactivation: *Bcl*I site in Em^r gene. Additional unique cloning sites: *Hind*III, *Cla*I.	17
pFX1	pDI25	5.5	Cm	−	No marker inactivation, unique cloning sites: *Hind*III, *Pvu*II, *Hind*III, *Hpa*II, *Mbo*I.	18
pFX3	pDI25	4.5	Cm	+	Contains α*lacZ* gene fragment for α-complementation in *E. coli*.	19

^a A low copy variant of pNZ18/19.

of resistance to chloramphenicol is an indication of the promoter strength, but for a more quantitative assay CAT activity can be determined according to the method of Shaw (20) in cell-free extracts (*Protocol 4*). A typical *cat-86* based promoter screening vector is shown in *Figure 4*. With the aid of this vector a number of promoter-carrying DNA fragments was obtained, sequenced, and mapped (13, 21). The way this vector is used is described in *Protocol 3*.

A promoter screening vector containing the promoterless *E. coli* β-galactosidase gene (*lacZ*) as the probe gene is pORI13 (J. W. Sanders and K. J. Leenhouts, unpublished data; *Figure 5*). In this vector the RBS and start codon of a lactococcal gene (the putative gene encoding fructose 1,6 biphosphate aldolase) was precisely fused to the coding frame of the *lacZ* gene. Upstream of the RBS an MCS has been inserted and this pWV01-derived vector also contains the Em[r] marker. The plasmid lacks the pWV01 *repA* gene and can therefore be used both as an integration vector and a replicating vector. Replication requires special strains which will be described in Section 2.2.3. The strength of promoters cloned into this vector can be determined by measuring in cell-free extracts (*Protocol 4*) the β-galactosidase activity on the substrate ONPG according to the method of Miller (22).

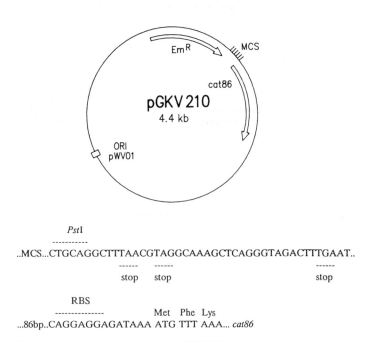

Figure 4. The promoter screening vector pGKV210. Unique restriction enzyme recognition sites in MCS: *Eco*RI, *Sma*I, *Bam*HI, *Sal*I, *Pst*I. Translational stop codons in all three reading frames are indicated. RBS: ribosome-binding site.

Protocol 3. Use of the promoter screening vector pGKV210

Materials

- restriction enzyme(s)
- T4 DNA ligase

- GM17 agar plates containing increasing concentrations of chloramphenicol
- sterile toothpicks

Method

1. Digest pGKV210 with one of the enzymes of the MCS (*Figure 4*).

2. Digest total DNA of the desired lactococcal strain with a restriction enzyme which generates termini compatible with the linearized pGKV210 DNA.

3. Ligate restriction fragments of the total DNA with pGKV210 DNA using standard conditions.

4. Transfer 1–2 µl of the ligation mixture to a plasmid-free *L. lactis*[a] strain by electrotransformation (*Protocol 1*). Plate the cells on GSM17 agar plates containing erythromycin (5 µg/ml) and chloramphenicol (4 µg/ml).

5. Transfer colonies using toothpicks to GM17 agar plates containing increasing concentrations of chloramphenicol (e.g. 5, 10, 15, 20, 25 µg/ml, etc.).

6. Make cell-free extracts (*Protocol 4*) and determine promoter strength using a CAT activity assay (20).

[a] The ligation mixture can also be transferred to *E. coli* or *B. subtilis*. For selection in *E. coli* use media containing 100 µl/ml erythromycin and 150 µg/ml chloramphenicol, and subsequently transfer colonies to plates containing increasing concentrations of chloramphenicol (150, 200, 250, 300 µg/ml, etc.). For selection in *B. subtilis* use media containing 5 µg/ml erythromycin and 20 µg/ml chloramphenicol, and transfer colonies to plates containing increasing concentrations of chloramphenicol (20, 30, 40, 50, 60 µg/ml, etc.).

Protocol 4. Preparation of cell-free lactococcal extracts

Materials

- GM17: M17 (37.25 g/litre; Difco, Detroit, USA), 0.5% (w/v) glucose
- demineralized water
- 25 ml (or larger) tubes
- bench-top centrifuge

- microcentrifuge
- glass beads
- mini bead-beater
- sterile pipette tips, 2 ml screw-cap microcentrifuge tubes, and 2 ml plastic tubes

Method

1. Inoculate the strain in 25 ml GM17 containing selective antibiotics, grow the culture at 30°C.

Protocol 4. *Continued*

2. Pellet the cells of the total culture volume by centrifugation (5000 r.p.m., 5 min).

3. Resuspend the cells in 10 ml demineralized water and centrifuge as in step **2**.

4. Resuspend the cells in 1 ml of an appropriate buffer[a] and transfer the solution to a 2 ml screw-cap microcentrifuge tube.

5. Add glass beads until the tube is almost filled.

6. Disrupt the cells in a mini bead-beater (a modified sanding machine) for 5 min at 4°C.

7. Chill the cell extract on ice.

8. Spin down glass beads and cell debris by centrifugation (10 000 r.p.m., 5 min).

9. Transfer supernatant to a clean 2 ml tube and store on ice.

[a] The composition of the buffer depends on the enzyme assay to be carried out. Optimal buffer for the β-galactosidase assay (m): 50 mM Na phosphate pH 7.0. Optimal buffer for the phospho-β-galactosidase assay: 50 mM Na phosphate pH 7.0 + 1 mM DTT. Optimal buffer for a CAT assay: 100 mM Tris pH 7.8 + 0.1 mM acetyl-CoA + 0.4 mg/ml DTNB (5,5'-Dithiobis-2-nitrobenzoic acid).

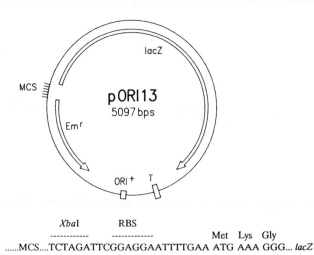

Figure 5. Plasmid pORI13, a vector suitable for promoter screening and for the construction of chromosomal transcriptional fusions. Ori[+]: plus origin of replication of the lactococcal plasmid pWV01, lacking the *repA* gene (see also Section 2.2.3.ii). Unique restriction enzyme recognition sites in MCS: *BgI*II, *Sph*I, *Nco*I, *Sma*I, *Eco*RI, *Asu*II, *Pst*I, *Sal*I, *Bam*HI, *Not*I, *Xma*III, *Xba*I. Translational stop codons in all three reading frames are indicated. RBS: ribosome-binding site. T: transcriptional terminator of *prtP*.

The probe genes in the promoter screening vectors described above are of heterologous origin. However, the screening vector pNZ336 incorporates the promoterless gene for lactococcal phospho-β-galactosidase (*lacG*), in addition to the transformation selection markers Em[r] and Cm[r] (23). The vector contains a MCS upstream of the RBS of the *lacG* gene (*Figure 6*). The strength of the promoters cloned into this vector can be determined by measuring in cell-free extracts (*Protocol 4*) the phospho-β-galactosidase activity on the substrate pONPG, according to Miller (22).

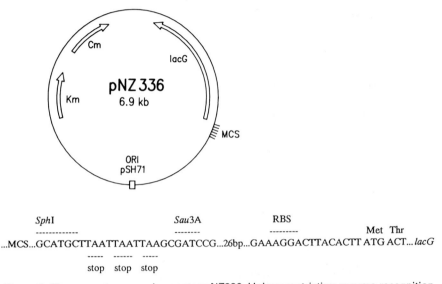

Met Thr

...MCS...GCATGCTTAATTAATTAAGCGATCCG...26bp...GAAAGGACTTACACTT ATG ACT...*lacG*

Figure 6. The promoter screening vector pNZ336. Unique restriction enzyme recognition sites in MCS: *Sal*I, *Hpa*I, *Cla*I, *Bam*HI, *Sma*I, *Kpn*I, *Sph*I. Translational stop codons in all three reading frames are indicated. RBS: ribosome-binding site.

ii. Vectors for assaying transcription terminator activity

Promoter screening vectors are also useful for selecting transcriptional terminators or for assaying the functionality of (presumed) terminators in lactococci. This can be readily achieved, especially in promoter screening vectors containing an MCS. A segment of DNA having promoter activity is first cloned in one of the restriction sites of the MCS, and the presumed termination structure is then inserted in a different, downstream, site of the MCS. Two such types of vector have been used for determining terminator strength in lactococci: pGKV21, in which transcription of the *cat-86* gene is driven by a promoter from the *Bacillus subtilis* bacteriophage SPO2 (13), and pGKV259, in which the transcription of the *cat* genes is driven by the strong lactococcal promoter P59 (21).

2.2.2 Gene expression vectors

Since the characterization of a number of *Lactococcus*-specific promoters and lactococcal genes several expression vectors have been developed for both constitutive and inducible expression of cloned genes.

i. Constitutive gene expression vectors

An important gene in lactococci is that which specifies the cell envelope-associated proteinase, as it is responsible for the initial degradation of milk caseins. This gene, *prtP*, has been cloned and sequenced from different lactococcal species. It is transcribed from a divergent promoter region that also drives transcription of a proteinase maturation gene, *prtM*. The gene product is required for activation the proteinase. Although both the *prtP* and the *prtM* promoters have been used to drive the expression of heterologous genes, the plasmids used did not incorporate all the features necessary for easy insertion of the genes to be expressed. In addition, the *prtP* and *prtM* promoters are relatively weak.

By means of promoter screening a number of lactococcal promoters have been isolated and characterized, some of which have been used for the construction of multipurpose expression vectors. The strong promoter P32 drives transcription in plasmid pMG36e (24), which incorporates the MCS of pUC18 and the transcriptional terminator of the *prtP* gene. P32 is located on a DNA segment that also contains the 5′ part of the open reading frame of the putative fructose 1,6 biphosphate aldolase gene of *Lactococcus*, and both in-frame and out-of-frame fusions can be made with this open reading frame (see *Figure 7*). This vector has been used for the production in lactococci of a variety of lysozymes (24, 25) and the *Bacillus subtilis* neutral protease, which is initially produced as a pre-pro-enzyme that is faithfully processed by lacto-cocci during secretion (26). As the P32 promoter is flanked by unique restriction sites, it can be easily replaced by other promoters. In pMG36e the effect of translational coupling has been investigated to optimize expression of β-galactosidase, by introducing a stop codon in the coding part of P32 and positioning the start codon of β-galactosidase at various distances from that stop codon. By making the stop and start codons partially overlapping, expression of β-galactosidase in *Lactococcus lactis* strain IL1403 was increased almost three-fold (27).

ii. Inducible expression vectors

Because high level expression of heterologous genes may be deleterious to the host cells, regulated gene expression signals have been incorporated in some expression vectors. One such inducible promoter is the *L. lactis dnaJ* promoter which has been used to over-express the α-amylase of *B. stearother-mophilus* by heat induction (M. Van Asseldonk, A, Simons, H. Visser, W. M. De Vos, and G. Simons, in preparation). Another is a versatile system based on the inducibility of the *L. lactis lac*-promoter, the *E. coli* T7 RNA

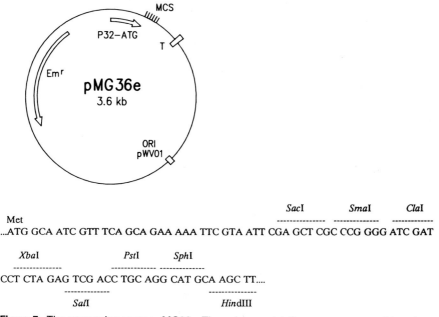

Figure 7. The expression vector pMG36e. The unique restriction enzyme recognition sites in the MCS are indicated. T: transcriptional terminator of *prtP*.

polymerase and the T7 promoter that was constructed to facilitate the over-production of any gene of interest by induction with lactose. This system depends on the presence of three compatible plasmids in one strain. The first plasmid carries the *L. lactis lac*-operon that is required for growth on lactose and for the production of the inducer. The second plasmid contains the *lacR* repressor and the *E. coli* T7 RNA polymerase gene under transcriptional control of the *lac*-promoter. In this configuration the T7 RNA polymerase is repressed during growth on glucose, but is expressed during growth on lactose. The T7 RNA polymerase then acts on the T7 promoter, located on the third plasmid (designated pLET), and drives expression of the gene of interest cloned downstream from the promoter. Several pLET plasmids have been constructed with convenient restriction sites to insert the target gene. The vector pLET2, shown in *Figure 8*, contains DNA sequences encoding the signal peptide of the lactococcal secreted protein Usp45 and has been successfully used to over-produce and to secrete the tetanus toxin fragment C (28, J. H. Wells, personal communication).

iii. Secretion and protein export signal selection vectors

At present four extracellular lactococcal proteins are known: PrtP, PrtM, OppA, and Usp45, an *u*nidentified *s*ecreted *p*rotein. All these proteins are initially synthesized as precursor molecules that carry a typical Gram-positive signal peptide that is cleaved or lipomodified during secretion. The signal sequence of PrtP has been used to direct the export of bovine prochymosin, the milk-clotting enzyme (29). The signal sequence of Usp45 has been used to direct the export of α-amylase from *B. stearothermophilus* (30; see also Section 2.2.2.*ii*) and the tetanus toxin fragment C (J. H. Wells, personal communication; see also Section 2.2.2.*ii*). As more efficient protein export signals than the ones currently in use may be present on the lactococcal chromosome an export signal selection vector was developed, in which the reporter gene is the α-amylase gene of *B. licheniformis* from which the native signal peptide-encoding DNA had been removed. In this promiscuous pWV01-based vector, shown in *Figure 9*, transcription in pGA14 is driven by the *B. subtilis* bacteriophage promoter SPO1. Signal sequence-encoding DNA, devoid of a promoter, can be selected with this vector. By using pGA14 several protein export functions have been isolated in *E. coli*, that were also active in *B. subtilis* and *L. lactis*. One of the signal peptides (AL9) allowed the growth of *Lactococcus* on starch as the sole carbon source (31). This characteristic allows pGA14 to be used for the direct selection of signal sequences in *L. lactis*. After cloning DNA fragments into the MCS of pGA14 and transforming the ligation mixture into *L. lactis*, the recombinant colonies can be screened for the production and secretion of α-amylase on selective agar plates containing starch (1%) by flooding the plates with an iodine solution. Colonies which actively secrete α-amylased produce a clear halo.

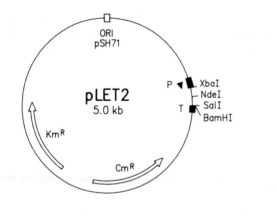

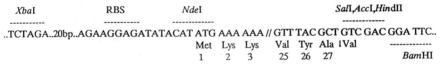

*Xba*I	RBS	*Nde*I						*Sal*I,*Acc*I,*Hind*II

..TCTAGA..20bp..AGAAGGAGATATACAT ATG AAA AAA // GTT TAC GCT GTC GAC GGA TTC..

                               Met Lys Lys   Val Tyr Ala ∣Val

                                1   2   3   25 26 27   *Bam*HI

Figure 8. The expression-secretion vector pLET2. Restriction enzyme recognition sites for obtaining translational fusions are indicated. RBS: ribosome-binding site. 1–27: signal peptide Usp45. The *arrow* indicates the recognition site for the signal peptidase. P▷: T7 promoter. T: transcriptional terminator bacteriophage T7 gene 10.

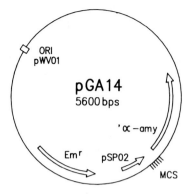

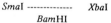

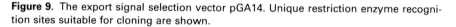

```
                 Gly   Asp Pro Leu Glu Ser Thr Ala Ala   Ala  Ala
..GAATTCGAGCTCGCC CGG GAT CCT CTA GAG TCG ACC GCA GCG GCG GCA .. α-amy
                 --------------      ----------------
                 SmaI --------------  XbaI
                      BamHI
```

Figure 9. The export signal selection vector pGA14. Unique restriction enzyme recognition sites suitable for cloning are shown.

2.2.3 Integration vectors

A surprisingly large number of genes relevant for dairying are located on plasmids in lactococci (for review see 36) with the inherent disadvantage that such genes can be lost because of segregational plasmid instability. This has been the primary reason why geneticists have screened a number of plasmids for potential use as integration vectors for lactococci. Two types of such vectors have been developed: those derived from heterologous plasmids unable to replicate in lactococci (*Table 4*) and a class derived from the lactococcal replicon pWV01 (*Table 3*).

i. Integration vectors based on heterologous plasmids
Although heterologous plasmids are not suitable to stabilize genes in lactococci for use in dairy practice, they have been used to study the feasibility of plasmid integration in *Lactococcus*. These plasmids usually have a lactococcal chromosomal fragment to provide homology for a homologous recombination event (*Table 4*). However, plasmids pTRK74 and pTRK145 integrate into the lactococcal chromosome by means of cointegrate formation with the chromosome at random locations (37). This property makes these vectors potentially useful for random mutagenesis strategies.

Stabilization by means of integration into the chromosome of *L. lactis* strain MG1363 was successfully achieved for the *prtP–prtM* genes, which play a key role in the lactococcal proteolytic system, by using pKL400B as the integration vector (38).

Table 3. Special purpose vectors based on small cryptic lactococcal replicons

Vector	Replicon	Size (kb)	Markers	MCS	Remarks	Reference
Promoter screening						
pGKV210	pWV01	4.4	Em	+	Contains promoterless *cat-86*.	13
pBV5030[a]	pWV01	4.3	Em	+	Contains promoterless *cat-86*ΔAC1.	32
pORI13[b]	pWV01(Ori+)	5.1	Em	+	Contains promoterless *lacZ*.	(J. W. Sanders unpublished data)
pNZ336	pSH71	6.9	Em Cm	+	Contains promoterless *p-β-gal*.	23
Terminator screening						
pGKV21	pWV01	4.9	Em Cm	+	Contains between SPO2 promoter and Cm^r gene unique *Bam*HI, *Pst*I, and *Sal*I sites.	13
pGKV259	pWV01	5.0	Em Cm	+	Contains between P59 promoter and Cm^r gene unique *Sal*I and *Pst*I sites.	21
Translation fusion						
pFX4, 5, and 6	pDI25	6.7	Cm	+	Suitable for translational fusions with *lacZ*.	19
pMG14	pWV01	7.7	Em	+	Suitable for translational fusions with *lacZ*.	33
pGKH10	pWV01	9.4	Em	−	Suitable for translational fusions in the unique *Bam*HI site with either *lacZ*, or α-*gal*, or with both in the case of bidirectional expression signals.	34
Expression and secretion						
pMG36e	pWV01	3.6	Em	+	Constitutive expression vector containing promoter P32.	24
pLET1[c]	pSH71	5.0	Cm Km	+	Inducible expression vector containing promoter T7.	(J. H. Wells, personal communication)
pLET2[c]	pSH71	5.0	Cm Km	+	Inducible expression-secretion vector containing promoter T7 and signal sequence of *usp45*.	(J. H. Wells, personal communication)
pLET3[c]	pSH71	5.0	Cm Km	+	Inducible expression-secretion vector containing promoter T7 and signal sequence of *prtP*.	(J. H. Wells, personal communication)

Signal sequence selection

Name	Replicon	Size (kb)	Marker		Description	Reference
pGA14	pWV01	5.6	Em	+	SPO2 promoter and *amy* gene devoid of signal sequence.	31
pGB14	pWV01	5.9	Em	+	Promoterless β-lactamase gene devoid of signal sequence and promoter.	31

Integration

Name	Replicon	Size (kb)	Marker		Description	Reference
pORI28[b]	pWV01(Ori$^+$)	1.7	Em	+	Suitable for Campbell-type integration into chromosomal location of choice.	(K. J. Leenhouts, unpublished data)
pINT23[b]	pWV01(Ori$^+$)	2.5	Em	+	Suitable for Campbell-type integration into *pepX* gene, inactivating *pepX*.	(K. J. Leenhouts, unpublished data)
pORI280[b]	pWV01(Ori$^+$)	5.3	Em lacZ	+	Suitable for replacement integration into chromosomal location of choice.	(K. J. Leenhouts, unpublished data)
pINT29[b]	pWV01(Ori$^+$)	4.7	Em	+	Suitable for replacement integration into *pepX* gene; replacement inactivates *pepX*, but Campbell-type integrations do not inactivate *pepX*.	(K. J. Leenhouts, unpublished data)
pORI13[b]	pWV01(Ori$^+$)	5.1	Em	+	Suitable to generate transcriptional fusions in the chromosome by Campbell-type integration.	(J. W. Sanders unpublished data)
pORI19[b]	pWV01(Ori$^+$)	2.2	Em	+	Suitable for random integrations into the chromosome by Campbell-type integration.	(J. Law unpublished data)
pG$^+$host4[d]	pWV01(Ts)	3.8	Em	+	Suitable for both Campbell-type and replacement recombination in the chromosome.	35
pG$^+$host5[d]	pWV01(Ts) pBR322	5.3	Em	+	As pG$^+$host4, in addition capable of replication in *E. coli* at 37°C.	(A. Gruss, unpublished data)

[a] Allows selection of less strong promoters, due to a four times higher CAT activity.
[b] Requires special Rep$^+$ *E. coli*, *B. subtilis*, or *L. lactis* helper strains for replication, see Section 2.2.3.ii.
[c] Requires special *L. lactis* strain containing a plasmid with the *L. lactis lac*-operon and a plasmid containing the T7 RNA polymerase gene, see Section 2.2.2.ii.
[d] Replicates at 28°C and integrates at 37.5°C in *L. lactis*, see Section 2.2.3.ii.

Table 4. Integration plasmids of heterologous origin that have been successfully used for integration into the chromosome of lactococci

Vector	Derived from plasmid	Size (kb)	Antibiotic resistance marker	Remarks	Reference
pHV60A/B[a]	pBR322	6.8	Cm	Some degree of illegitimate integration. Plasmid amplification gradually lost under non-selective conditions; one plasmid copy retained.	39
pKL10A/B[a]	pBR322	3.6	Em	Plasmid amplification probably by integration of plasmid multimers.	40
pKL203A/B[a]	pUB110	6.5	Em Cm	Unstable plasmid amplification.	40
pKL400B[a]	pTB19	7.7	Em	Stable single-copy plasmid integration.	40
pKL301B[a]	pSC101	9.6	Cm	Stable single-copy plasmid integration.	40
pML336	pUC18	9.1	Em	Stable single and double cross-over integration in *pepX* gene.	41
pE194/φ	pE194	—	Em	Plasmid pE194 containing a fragment of a temperate *Lactococcus* phage; the host (IL1403) was lysogenic for the phage. Stable single and double cross-over integration.	42
pTRK74[b]	p15A	7.3	Tc Em Cm	Contains the lactococcal IS element IS946; integration by random cointegrate formation with the chromosome.	37
pTRK145[b]	p15A	5.7	Em Cm	As pTRK74.	37

[a] A and B refer to two randomly chosen chromosomal fragments of approximately 1.3 kb of strain MG1363.
[b] Potentially useful for random mutagenesis of the chromosome (see Section 2.2.3.*i*).

ii. Integration vectors based on a lactococcal replicon

An important stimulus for the development of integration systems based on a lactococcal replicon has been the consideration that only recombinant DNA-modified lactococci may be used for food production if the vehicles are completely composed of lactic acid bacterial DNA. Apart from this practical motivation there was also a need for the development of systems to analyse the genetics of the lactococcal chromosome and, consequently, vectors have recently been developed that allow:

- the insertional inactivation of random lactococcal genes
- the construction of transcriptional fusions in the lactococcal chromosome (*Table 3*). Some of these vectors are discussed below.

Campbell-type integration vectors

Two systems for plasmid integration into the lactococcal chromosome have been developed based on pWV01-derived vectors. In one of these the pWV01 derivative was made incapable of replication in lactococci, by deleting the *repA* gene essential for replication, while the plus origin of replication, which is the site of replication initiation, was retained (43; Ori$^+$-system). In the other system, the *repA* gene was mutated such that the plasmid became temperature-sensitive for replication (35; Ts-system). In the Ori$^+$-system plasmids lacking the *repA* gene are multiplied in *L. lactis, B. subtilis,* or *E. coli* strains in which the *repA* gene, under the control of the strong lactococcal promoter P23, is integrated (Rep$^+$ helper strains). After extraction from the Rep$^+$ helper strains, the integration vector carrying a lactococcal chromosomal DNA fragment is subsequently introduced in *rep* lactococci (Rep$^-$ strains, e.g. strain MG1363 or IL1403) in which the vector is integrated by means of a Campbell-type recombination owing to the presence of a lactococcal chromosomal DNA fragment. Thus, in this system the Rep$^+$ helper strains are the source of the integration plasmid. A diagram of the procedure is shown in *Figure 10*.

An example of such an integration plasmid is pINT23 (*Figure 11*). This plasmid carries an internal fragment of the lactococcal *pepX* gene, a gene which specifies the X-prolyl dipeptidyl aminopeptidase. Integration of the plasmid via this chromosomal fragment results in the inactivation of this non-essential gene. This event is readily scorable because mutant colonies remain white in an assay procedure using H-Gly-Pro-βNA and Fast Garnet GBC Salt, whereas wild-type colonies stain red (*Protocol 6*). The procedure to use pINT23 as integration vector is described in *Protocol 5*.

Protocol 5. Use of the integration vector pINT23

Materials

- pINT23 plasmid DNA
- one of the Rep$^+$ *L. lactis, B. subtilis,* or *E. coli* helper strains
- Rep$^-$ *L. lactis* strain, e.g. MG1363 or IL1403

Method

1. Clone the DNA fragment of interest in the MCS of pINT23 (see *Figure 11*).

2. Transform the ligation mixture into one of the Rep$^+$ *L. lactis* (*Protocol 1*), *B. subtilis,* or *E. coli* helper strains (see *Figure 10*).

3. Isolate plasmid DNA of the correct construction from the Rep$^+$ helper strain.

Protocol 5. Continued

4. Transform Rep⁻ *L. lactis* strain (e.g. MG1363 or IL1403) with 0.1–1 µg plasmid DNA (*Protocol 1*). Usually, 300–400 transformants per µg DNA are obtained.

5. Perform PepX assay (*Protocol 6*) to verify that integration occurred by homology.

Protocol 6. PepX plate assay

Materials

- 0.5% agarose in demineralized water (keep at 60°C)
- 0.2 M Tris pH 7.4
- Fast Garnet GBC Salt (Sigma, St. Louis, USA)
- DMF (dimethylformamide)
- H-Gly-Pro-βNA: the dipeptide glycine-proline coupled to the chromogenic substrate β-naphthylamide (BaChem, Bubendorf, Switzerland)
- GM17 agar plates and/or GM17 liquid medium
- sterile toothpicks

Method

1. Incubate agar plate to form colonies for at least 16 h at 30°C.[a]

2. Flood plates (9 cm diameter) with 3 ml 0.5% agarose, such that all colonies are covered. Wait until the agarose has solidified.

3. Mix 5 ml 0.2 M Tris pH 7.4 containing Fast Garnet GBC Salt (2 mg/ml) with 0.2 ml DMF containing H-Gly-Pro-βNA (10 mg/ml).

4. Pour the mixture on the plate and incubate at room temperature for 20 min or longer.[b]

5. Rinse the plates with water and determine the phenotype of the colonies:

   - PepX⁺ colonies stain red
   - PepX⁻ colonies remain white

6. Transfer with a toothpick the colony of interest to a fresh agar plate or liquid medium for further analysis.

---

[a] Longer incubation time improves intensity of the staining reaction.
[b] Longer incubation times improve staining, but affect cell survival.

The Ts-system for chromosomal plasmid integration exploits the temperature-sensitivity of a pWV01-derived vector. This vector, pH⁺host4 (*Figure 12*), carrying four mutations in the *repA* gene, is fully capable of replication in *L. lactis* at 28°C, but is unable to do so at 37.5°C. A useful derivative for cloning in *E. coli* that is able to replicate at 37°C was obtained by a fusion of pG⁺

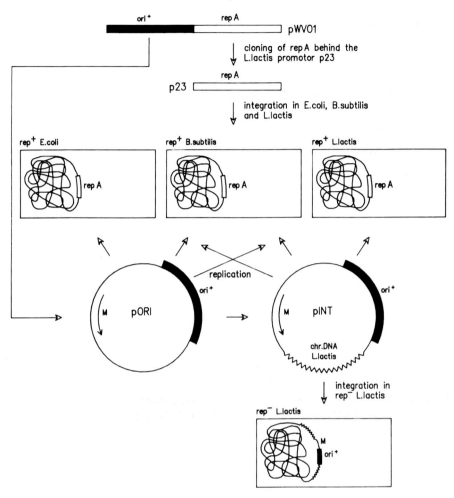

Figure 10. Schematic representation of the Ori+-system. Ori+: fragment containing the plus origin of replication of pWV01. repA: fragment encoding the replication initiation protein of pWV01. rep+: RepA-producing strain. P23: lactococcal promoter, also active in *E. coli* and *B. subtilis*. pORI: Ori+-vector that does not contain lactococcal chromosomal DNA. pINT: Ori+-vector containing lactococcal chromosomal DNA. rep−: strain not producing RepA. chr. DNA: chromosomal DNA. M: selectable marker.

host4 with the origin of replication of pBR322, resulting in pG+host5. The advantage of using these vectors is that high frequencies of integration can be obtained, by simply shifting the culture to the non-permissive temperature and subsequent selection for antibiotic resistance. Chromosomal fragments as small as approximately 300 bp can be used to obtain integration (T. Biswas, A. Gruss, D. Ehrlich and and E. Maguin, in preparation). The plasmid con-

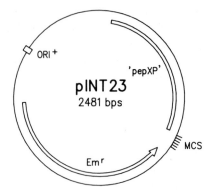

Figure 11. The Ori⁺-integration plasmid pINT23. Unique cloning sites in MCS: *Bg*lII, *Cla*I, *Sph*I, *Nco*I, *Kpn*I, *Sma*I, *Sac*I, *Eco*RI, *Asu*II, *Pst*I, *Aat*II, *Nde*I, *Eco*RV, *Bg*lI, *Not*I, *Xma*III, *Xba*I. 'pepXP': internal fragment of the *pepX* gene. For use of the plasmid see *Protocol 5*.

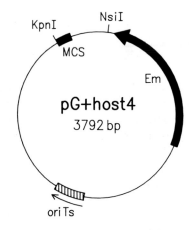

Figure 12. The temperature-sensitive plasmid pG⁺host4. Unique cloning sites in MCS: *Kpn*I, *Apa*I, *Xho*I, *Sal*I, *Cla*I, *Hind*III, *Eco*RV, *Eco*RI, *Pst*I, *Sma*I, *Spe*I, *Xba*I, *Not*I, *Sac*II, *Bst*XI. ori Ts: temperature-sensitive origin of replication of pWV01. For use of the plasmid: see Section 2.2.3.ii.

taining the chromosomal insert is used to transform the appropriate lactococcal strain; a high efficiency of transformation is not required, as integration is performed as a separate step. Once the strain containing the replicating integration plasmid is established (*Protocol 1*) and verified (*Protocol 2*), it is grown overnight at 28°C in the presence of erythromycin. Cells are then diluted 100-fold in this medium and grown at 28°C for 2 to 2.5 h (to log phase). Cultures are then shifted to 37.5°C for 3 h (between six and nine generations) to lower the plasmid copy number per cell. Samples are diluted and plated at 37°C on GM17Em plates to detect the integration event. For subsequent use, the integrants isolated at 37°C are routinely maintained at 37.5°C in selective media to prevent plasmid excision. Southern hybridization and biological tests (where possible) should be used to verify that integration occurred by homology.

Vectors for random insertional gene inactivation and production of gene probes

The pWV01-based integration systems described above can also be used to obtain random gene interruptions. A bank of small chromosomal DNA fragments is constructed in Ori$^+$ or Ts-plasmids in suitable *L. lactis, B. subtilis*, or *E. coli* strains and the bank is subsequently transferred to the target host. This procedure will result in a collection of transformants with plasmids integrated at random sites in their chromosomes. The collection of integrants can then be screened for mutant phenotypes. Owing to the small size of the chromosomal inserts, the probability that a gene will be interrupted by the integrated plasmid is high. Once a mutant colony has been identified, parts of the mutated gene can be cloned by cutting the chromosomal DNA of the mutant with restriction enzymes which have only one or no recognition sites in the integrated vector. The restricted DNA is subsequently diluted, religated, and transferred to a host which allows replication of the vector (in the case of the Ori$^+$-system a Rep$^+$ helper strain should be used). The transformants will carry part(s) of the interrupted gene that can subsequently be used for screening purposes to isolate the complete gene. An important advantage of using a pWV01-derived integration vector instead of a heterologous integration vector is that the homologous host (*L. lactis*) can be used to clone the gene. This procedure minimizes deletion formation in DNA sequences that are difficult to clone in non-compatible backgrounds.

An important step in the procedure is the construction of a plasmid bank containing random chromosomal inserts. To optimize this step a vector was constructed in the Ori$^+$-system which allows efficient screening for inserts. The Ori$^+$-vector, pORI19 (*Figure 13*), carries the α*lacZ* gene fragment containing the MCS of pUC19, which allows the use of the α-complementation system for screening of recombinant vectors in the Rep$^+$ *E. coli* helper strain (EC101: strain JM101 containing an integrated copy of the lactococcal *repA* gene). This system has been used to construct, in the Rep$^+$ *E. coli*, plasmid banks with *Sau*3A, *Alu*I, or *Rsa*I DNA fragments of the MG1363 chromosome (obtained with different partial digests; average insert size 600 bp). Transfer of these plasmid banks to strain MG1363 (*Protocol 1*) resulted in more than 1000 transformants per microgram of plasmid DNA. The transformants contained integrated plasmids at random sites in the chromosome (J. Law and K. J. Leenhouts, unpublished data).

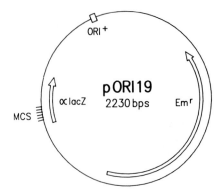

Figure 13. Plasmid pORI19, a vector suitable for random mutagenesis of the lactococcal chromosome. Unique cloning sites in MCS: *Eco*RI, *Sac*I, *Kpn*I, *Sma*I, *Bam*HI, *Xba*I, *Sal*I, *Acc*I, *Pst*I, *Sph*I, *Hin*dIII.

Vector for in vivo *production of transcriptional fusions*

To determine the promoter activity of chromosomally located promoters *in vivo* (i.e. in one copy in *L. lactis*), vector pORI13 was developed (*Figure 5*). This Ori⁺-vector can be used to make transcriptional fusions in the lactococcal chromosome by Campbell-type integration. Cloning of DNA fragments containing a promoter and/or the 5'-end of a gene result, after integration of the plasmid, in a transcriptional fusion with a gene in addition to a wild-type copy of that gene. In contrast, cloning of internal gene fragments into pORI13 results, after integration in the chromosome, in a transcriptional fusion with a gene, in addition to a mutated copy of that gene. The promoter activity is subsequently determined by measuring the β-galactosidase activity on ONPG in cell-free extracts according to the protocol of Miller (22) (*Protocol 4*).

Systems for replacement recombination

In replacement recombination, genetic information usually contained in a plasmid is introduced into the chromosome by homologous recombination involving two cross-over events. If the plasmid, which should again be of the type that can not replicate in lactococci, contains a chromosomal gene carrying a deletion, the double cross-over produces the equivalent deletion in the chromosome. If the gene carries an insertion (for example, another gene of interest) the insertion will be added to the host chromosome.

Recently, two systems were developed to allow the easy selection of replacement recombination in lactococci. One of these systems employs the pWV01-derived Ori⁺-system and the other uses the pWV01-derived Ts-replication system. The Ts-system makes use of the plasmids pG⁺ host 4 or 5. The Ori⁺-system uses the derivative pORI280, which contains the *E. coli lacZ*

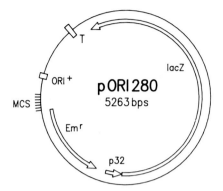

Figure 14. Plasmid pORI280, a vector suitable for obtaining replacement integration. Unique cloning sites in MCS: *Bgl*II, *Sph*I, *Nco*I, *Sma*I, *Asu*II, *Eco*RI, *Hind*III, *Pst*I, *Mlu*I, *Aat*II, *Nde*I, *Bam*HI, *Bgl*I, *Not*I, *Xma*III, *Xba*I. T: transcription terminator of *prtP*. p32: lactococcal promoter also active in *E. coli* and *B. subtilis*.

gene under transcriptional control of the lactococcal promoter p32 (*Figure 14*). The strategy for obtaining replacement recombination is outlined in *Figure 15*. The starting plasmid used is either composed of a pORI280 or a pG$^+$ host vector carrying a chromosomal segment (A + B in *Figure 15*) interrupted by an insertion (e.g. a gene of interest) or any other modification (e.g. point mutation or deletion). In the Ts-system the plasmid is first established in the replicating state in the strain and maintained at 28°C with selection. The plasmid is subsequently integrated (step I in *Figure 15*), via segment A or B by a temperature shift, followed by a plating procedure as described in Section 2.2.3*ii*. The pORI280 construct is directly integrated, via segment A or B, after the introduction into the strain of interest. Transformants obtained with pORI280 derivatives stain blue on GM17 agar plates containing X-gal. Excision of the integrated plasmids (step II in *Figure 15*) is established by recombination via the repeats A or B, resulting in the wild-type strain or in gene replacement. For step II the Ts-system makes use of the fact that replication between repeats in the chromosome is enhanced if an active replicon is located between these repeats (44). A strain carrying the integrated pG$^+$ host plasmid is grown overnight at 37.5°C in GM17 with erythromycin to obtain a population of integrants. The culture is diluted 1:10^5, in GM17 without antibiotic, and shifted to 28°C for 12 h or more. The cells are plated at 37°C on non-selective GM17 agar plates. The colonies are screened for Em sensitivity (the yield is usually between 1 and 40%).

The Ori$^+$-system makes use in step II of spontaneous recombination between repeats in the lactococcal chromosome, which has a frequency of approximately 5 × 10^6 per generation in the *pepX* gene region (41). The strain carrying the integrated pORI280 vector is grown overnight at 30°C in the presence of erythromycin. The culture is diluted to a cell density of one to

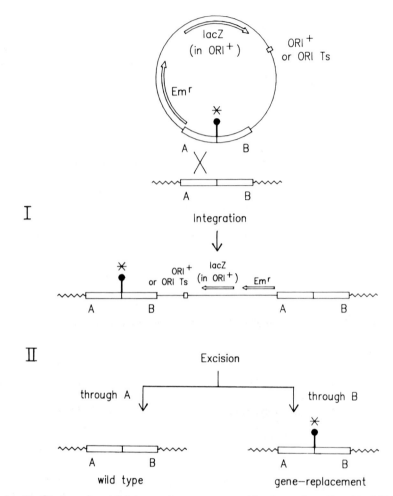

Figure 15. Strategy for obtaining replacement recombination using plasmid pORI280 or pG⁺host4 or 5. Ori⁺: fragment lacking *repA* but carrying the plus origin of replication of pWVO1. ori Ts: fragment carrying the gene encoding the ts RepA of pWVO1.A and B: two adjacent chromosome segments. The *asterisk* indicates a mutation (deletion or point mutation) or the insertion of a gene of interest.

ten cells per millilitre in GM17 without antibiotic and grown overnight at 30°C (∼ 30–35 generations). Approximately 10 000 cells are plated at 30°C on non-selective GM17 agar plates containing X-gal. Usually between one and ten white (Em^s) colonies develop on these plates. In both procedures Southern hybridization should be used to verify that replacement integration occurred. The absence of antibiotic resistance markers in the modified strain offers interesting possibilities towards the construction of food-grade modified lactococcal strains.

Acknowledgements

We would like to thank our colleagues W. M. de Vos, A. Gruss, M. Gasson, T. R. Klaenhammer, J. Law, J. W. Sanders, and also J. H. Wells, who kindly shared results prior to publication.

References

1. Gasson, M. J. (1983). *J. Bacteriol.*, **154**, 1.
2. Chopin, A., Chopin, M.-C., Moillo, A., and Langella, P. (1984). *Plasmid*, **11**, 260.
3. Holo, H. and Nes, I. F. (1989). *Appl. Environ. Microbiol.*, **55**, 3119.
4. Simon, D. and Chopin, A. (1988). *Biochimie*, **70**, 559.
5. Behnke, D., Gilmore, M. S., and Ferretti, J. (1981). *Mol. Gen. Genet.*, **182**, 414.
6. Macrina, F. L., Tobian, J. A., Jones, K. R., Evans, R. P., and Clewell, D. B. (1982). *Gene*, **19**, 345.
7. Dao, M. L. and Ferretti, J. J. (1985). *Appl. Environ. Microbiol.* 49, 115.
8. Achen, M. G., Davidson, B. E., and Hillier, A. J. (1986). *Gene*, **45**, 45.
9. Kok, J., Van der Vossen, J. M. B. M., and Venema, G. (1984). *Appl. Environ. Microbiol.*, **48**, 726.
10. Kiewiet, R., Kok, J., Seegers, J. F. M. L., Venema, G., and Bron, S. (1993). *Appl. Environ. Microbiol.*, **59**, 358.
11. De Vos, W. M. (1986). *Neth. Milk Dairy J.*, **40**, 141.
12. Gasson, M. J. and Anderson, P. H. (1985). *FEMS Microbiol. Lett.*, **30**, 193.
13. Van der Vossen, J. M. B. M., Kok, J., and Venema, G. (1985). *Appl. Environ. Microbiol.*, **50**, 540.
14. Birnboim, H. C. and Doly, J. (1979). *Nucleic Acids Res.*, **7**, 1513.
15. Ish-Horowicz, D. and Burke, F. J. (1981). *Nucleic Acids Res.*, **9**, 2989.
16. De Vos, W. M. and Simons, G. (1988). *Biochimie*, **70**, 461.
17. Von Wright, A., Tynkkynen, S., and Suominen, M. (1987). *Appl. Environ. Microbiol.*, **53**, 1584.
18. Xu, F., Pearce, L. E., and Yu, P. L. (1991). *Mol. Gen. Genet.*, **227**, 33.
19. Xu, F., Pearce, L. E., and Yu, P. L. (1991). *FEMS Microbiol. Lett.*, **77**, 55.
20. Shaw, W. V. (1975). *Methods in enzymology*, **43**, 737.
21. Van der Vossen, J. M. B. M., Van der Lelie, D., and Venema, G. (1987). *Appl. Environ. Microbiol.*, **53**, 2452.
22. Miller, J. (1972). *Experiments in Molecular Genetics* Cold Spring Harbor Laboratory Press, Cold Spring Harbor, New York.
23. Simons, G., Buys, H., Hogers, K., Koenhen, E., and De Vos, W. M. (1990). *Dev. Industr. Microbiol.*, **31**, 31.
24. Van de Guchte, M., Van der Vossen, J. M. B. M., Kok., J., and Venema, G. (1989). *Appl. Environ. Microbiol.*, **55**, 224.
25. Van de Guchte, M., Van der Wal, F. J., Kok, J., and Venema, G. (1992). *Appl. Microbiol. Biotechn.*, **37**, 216.
26. Van de Guchte, M., Kodde, J., Van der Vossen, J. M. B. M., Kok, J., and Venema, G. (1990). *Appl. Environ. Microbiol.*, **56**, 2606.
27. Van de Guchte, M., Kok, J., and Venema, G. (1991) *Mol. Gen. Genet.*, **227**, 65.

28. Wells, J. H., Wilson, D. W., Norton, P. M. N., Gasson, M. J., and Le Page, R. W. F. (1993). *Mol. Microbiol.*, (in press).
29. Simons, G., Rutten, G., Hornes, M., and De Vos, W. M. (1988). In *Proceedings of the 2nd Netherlands Biotechnology Congress* (ed. H. Breteler, D. H. van Lelieveld, and K. C. A. M. Luyben), p. 183. Netherlands Biotechnological Society, Zeist, The Netherlands.
30. Simons, G., Van Asseldonk, M., Rutten, G., Braks, A., Nijhuis, M., Hornes, M., and De Vos, W. M. (1990). *FEMS Microbiol. Rev.*, **87**, 24.
31. Perez Martinez, G., Kok, J., Venema, G., Van Dijl, J.M., Smith, H., and Bron, S. (1992). *Mol. Gen. Genet.*, **234**, 401.
32. Bojovic, B., Djordjevic, G., and Topisirovic, L. (1991). *Appl. Environ. Microbiol.*, **57**, 385.
33. Van de Guchte, M. (1991). Ph.D. Thesis, University of Groningen, Haren, The Netherlands.
34. Haandrikman, A. J. (1990). Ph.D. Thesis, University of Groningen, Haren, The Netherlands.
35. Maguin, E., Duwat, P., Hege, T., Ehrlich, D., and Gruss, A. (1992). *J. Bacteriol.*, **174**, 5633.
36. Venema, G. and Kok, J. (1987). *Trends Biotechnol.*, **5**, 144.
37. Romero, D. A. and Klaenhammer, T. R. (1992). *Appl. Environ. Microbiol.*, **58**, 699.
38. Leenhouts, K. J., Gietema, J., Kok, J., and Venema, G. (1991). *Appl. Environ. Microbiol.*, **57**, 2568.
39. Leenhouts, K. J., Kok, J., and Venema, G. (1989). *Appl. Environ. Microbiol.*, **55**, 394.
40. Leenhouts, K. J., Kok, J., and Venema, G. (1990). *Appl. Environ. Microbiol.*, **56**, 2726.
41. Leenhouts, K. J., Kok, J., and Venema, G. (1991). *J. Bacteriol.*, **173**, 4794.
42. Chopin, M. C., Chopin, A., Rouault, A., and Galleron, N. (1989). *Appl. Environ. Microbiol.*, **55**, 1769.
43. Leenhouts, K. J., Kok, J., and Venema, G. (1991). *Appl. Environ. Microbiol.*, **57**, 2562.
44. Noirot, P., Petit, M.-A., and Ehrlich, D. (1987). *J. Mol. Biol.*, **196**, 39.

4

Virulence plasmids

MARCELO E. TOLMASKY, LUIS A. ACTIS, and
JORGE H. CROSA

1. Introduction

The term plasmid was originally used by Lederberg (1) to describe all extra-chromosomal hereditary determinants. Currently, the term is restricted to autonomously replicating extrachromosomal DNA. Their sizes range from 1 kb to more than 200 kb (2), and even larger plasmids were detected in *Rhizobium* (3). Although they replicate autonomously, plasmids rely on host-encoded factors for their replication. Plasmids are not essential for the survival of bacteria, but they may nevertheless encode a wide variety of genetic determinants which permit their bacterial hosts to survive better in an adverse environment or to compete better with other micro-organisms occupying the same ecological niche. The medical importance of plasmids that encode for antibiotic resistance, as well as specific virulence traits has been well documented. In this chapter we describe techniques used in our laboratory for studying plasmid-encoded virulence factors such as the iron-uptake systems encoded by pJM1 in *Vibrio anguillarum* (4), and by pColV-K30 in *Escherichia coli* (5), as well as antibiotic resistance coded for by plasmids isolated from pathogenic *Klebsiella pneumoniae* strains (6).

2. Isolation of plasmids

In this section we describe some of the classical methods used to isolate plasmid DNA from a variety of bacterial species. There are several commercially available kits to isolate plasmid DNA.

2.1 Large scale method

2.1.1 Alkaline lysis and caesium chloride gradient centrifugation

The method described here is a shorter version of the one described by Birnboim and Doly (7). It combines cell lysis by treatment with alkaline sodium dodecyl sulfate (SDS) and high speed centrifugation in caesium chloride gradients containing ethidium bromide.

2.2.2 Acid phenol

Acid phenol extraction selectively removes nicked and linear DNA leaving only closed circular molecules (8). The following protocol (9) gives plasmid DNA preparations of a quality similar to that obtained after separation in a CsCl ethidium bromide gradient. This DNA can be used for sequencing and other procedures such as exonuclease III generated nested deletions.

Protocol 4. Small scale isolation of closed circular plasmid DNA

1. Proceed through step 4 as described in the small scale alkaline lysis method (*Protocol 2*). Transfer 400 µl of the supernatant to another microcentrifuge tube and add 400 µl of acid phenol (made by equilibration of distilled phenol with 0.05 M sodium acetate pH 4.0). Vortex vigorously and separate the phases in a microcentrifuge.

2. Transfer the upper phase to a microcentrifuge tube and extract with one volume of chloroform/isoamyl alcohol (24:1).

3. Transfer the upper phase to another microcentrifuge tube and add 2.5 volumes of cold 95% ethanol.

4. Recover the plasmid DNA by centrifugation for 5 min, pour off the supernatant, wash once with 70% cold ethanol, and dry. Resuspend the pellet in 20 µl of TE buffer containing 50 µg/ml of boiled RNase and incubate at 37°C for 30 min.

2.2.3 Hot Triton X-100 lysis method

This method was used in our laboratory for the analysis of plasmid profiles of *V. anguillarum* and clinical isolates of *A. baumanii* (10), with sizes ranging from 2 to 100 kb.

Protocol 5. Hot Triton X-100 lysis method

1. Grow overnight 40 ml culture with shaking in an appropriate rich medium in a 100 ml flask.

2. Harvest the cells by centrifugation at 3000 *g* for 10 min.

3. Suspend the cell pellet in 5 ml of TES buffer (50 mM Tris–HCl, 5 mM EDTA, 50 mM NaCl, pH 8.0).

4. Centrifuge the cells as before, and resuspend the cell pellet in 2 ml of a 25% sucrose solution (in 1 mM EDTA and 50 mM Tris–HCl, pH 8.0). Place the tube on ice for 20 min.

5. Add to the suspension 400 µl of lysozyme solution (10 mg/ml in 0.25 M Tris–HCl pH 8.0) keeping the tube on ice for another 20 min.

---

6. Add 800 μl of 0.5 M EDTA pH 8.0 to the cell suspension and lyse the cells by adding 4.4 ml of Triton lytic mixture (1 ml of 10% Triton X-100 in 10 mM Tris–HCl pH 8.0, 25 ml of 0.25 M EDTA pH 8.0, 5 ml of 1 M Tris–HCl pH 8.0, 69 ml of water). Mix gently.

7. Transfer the tube to a 65°C water-bath and incubate for 20 min.

8. Remove the cellular debris by centrifugation at 27 200 *g* for 40 min. Transfer the clear supernatant to a fresh centrifuge tube.

9. Adjust the solution to 0.5 M NaCl and 10% polyethylene glycol by use of stock solutions of 5 M NaCl and 40% polyethylene glycol (molecular weight 1000 to 6000).

10. Keep the tube overnight on ice.

11. Collect the precipitate by centrifugation at 3000 *g* for 10 min, and resuspend the pellet at 4°C in 1 to 2 ml of 0.25 M NaCl containing 1 mM EDTA and 10 mM Tris–HCl pH 8.0.

12. Add 2 volumes of 95% ethanol at −20°C, and let the tube stand overnight at −20°C. Alternatively, the DNA can be precipitated by allowing it to stand for 30 min at −70°C.

13. Spin down the DNA by centrifugation at 12 000 *g* at 4°C. Resuspend the DNA in TES.

3. Curing of plasmids

Curing of a plasmid is the best method to show that it encodes a virulence factor (11). The virulence phenotype linked to the presence of the plasmid will not be expressed in cured derivatives. Of course, reintroduction of the plasmid or a recombinant clone carrying the specific genetic trait must be accompanied by a regaining of virulence. Three plasmid curing protocols are described below.

Protocol 6. Curing using temperature

We describe here a protocol used to cure *V. anguillarum* (11, 12) which normally grows at 30°C. This protocol can easily be adapted to other bacteria. For example, we recently applied basically the same protocol to cure a plasmid from *Acinetobacter baumanii*, using L agar and broth (1% tryptone, 0.5% yeast extract, 0.5% NaCl, in the case of solid media 1.5% agar was added), and 42°C as the temperature of plasmid curing.

1. Inoculate 1 ml of Trypticase soy broth, supplemented with 1% NaCl, with 10 μl of an overnight culture of *V. anguillarum* (cultured at 30°C). Incubate at 37°C for 16 h.

Protocol 9. *Continued*

    plate (without any antibiotic) and incubate the plate overnight at 37°C. This step can also be carried out by centrifugation at 3000 *g* for 10 min. In this case the supernatant must be poured off and the cells can be placed on top of the plate using a sterile stick.

4. Plate the transconjugants on L agar plates containing Nal (to select only *E. coli* C2110), Ap (to select for the presence of Tn*3*-HoHo1), and the appropriate antibiotic for which the recombinant clone encodes resistance. In these conditions only *E. coli* C2210 harbouring the recombinant clone with a Tn*3*-HoHo1 insertion can grow. Incubate the plates overnight to obtain isolated colonies.

5. Prepare plasmid DNA from isolated colonies. The location of the transposon can be easily mapped by restriction endonuclease analysis, or if the DNA sequence of the clone is known, sequencing with a primer located at one end of Tn*3*-HoHo1 provides the exact location. Below (see *Protocol 14*) we describe a quick and easy sequencing protocol used to locate different types of mutations.

6. When the clones are from bacteria other than *E. coli*, the mutated recombinant plasmids can be introduced into the original genetic background for analysis of their phenotypes by carrying out another conjugation experiment.

4.2 *In vitro* chemical mutagenesis using hydroxylamine

This is a very convenient method to generate point mutations *in vitro*. Hydroxylamine produces a deamination of cytosine rendering GC to AT transitions. We used the following procedure (19) to mutate the pColV-K30 RepI protein gene (20). It is important to mention here that this mutagenesis is more useful when a positive selection for the mutants is available.

Protocol 10. Mutagenesis of plasmid DNA with hydroxylamine

1. In three tubes resuspend about 10 μg of purified plasmid DNA in 500 μl of freshly prepared hydroxylamine solution (90 mg NaOH, 350 mg hydroxylamine hydrochloride, 5 ml ice-cold water). Incubate the tubes at 37°C for 15, 20, and 25 h respectively.

2. After incubation, add 10 μl of 5 M NaCl, 50 μl of 1 mg/ml bovine serum albumin, and 1 ml of 95% ethanol. Pellet the DNA in a microcentrifuge.

3. Wash the DNA with 70% ethanol and resuspend in 400 μl of 0.3 M sodium acetate, pH 4.8. Reprecipitate by adding 880 μl of cold 95% ethanol and centrifugation. Wash the pellet with 70% ethanol, dry, and resuspend in 50 μl of TE buffer.

> **4.** Use this DNA to transform a suitable strain. Plate to obtain isolated colonies in the presence of selection for the desired phenotype. Isolate plasmid DNA (*Protocol 4*) and sequence to determine the location of the mutation.

4.3 Site-directed mutagenesis

Several methods have been developed to generate mutations at a specific location in a DNA molecule containing cloned genes (18). To generate point mutations the most popular method is the oligonucleotide-directed *in vitro* mutagenesis of single-stranded DNA. Since several commercial kits (Muta-Gene Phagemid in vitro Mutagenesis, Bio-Rad, Richmond, CA; Altered Sites in vitro Mutagenesis system, Promega, Madison, WI) exist to carry out this method we will not describe it here. Protocols vary for each kit, but they all consist basically of cloning the gene to be mutated in a phage or a phagemid, preparation of single-stranded DNA, phosphorylation of the oligonucleotide, hybridization to the single-stranded DNA, synthesis of the complementary strand by extending the oligonucleotide with DNA polymerase, transformation of a suitable bacterial strain, and analysis of the DNA to detect the mutants. We recently used this method with few modifications with double-stranded DNA (20). Other methods include site-directed insertion or deletion of DNA fragments.

4.3.1 Site-directed deletion or insertion mutagenesis

A particular fragment of cloned DNA can be mutagenized at a specific site by either the deletion or the insertion of a DNA fragment. The exonuclease *Bal*31 can be used to generate controlled DNA deletions on plasmids linearized with a restriction endonuclease that cuts the recombinant plasmid only once, within or near the sequence to be mutagenized. Alternatively, the exonuclease activity of the Klenow fragment of DNA polymerase I can be used, although in this case the deletion involves only a few bases and depends on the restriction enzyme used to linearize the plasmid. Site-directed insertion mutagenesis can be performed by using DNA fragments harbouring selectable markers such as antibiotic resistance, e.g. the Ω fragment present in the pHP45 plasmid (21) or the kanamycin resistance cassette of pUC4K (Pharmacia, Piscataway, NJ). We have used extensively in our work the Ω fragment which is very useful for mutagenesis since it carries at both ends transcription and translation termination signals in all reading frames, besides resistance to streptomycin and spectinomycin as selectable markers. These features make this fragment ideal for effective gene interruption, since they avoid transcriptional or translational fusions.

electrophoresis conditions we used the plates were hot enough to provide good separation.

(b) Power supply that can provide 1000 V.

(c) Gel dryer.

Protocol 14. Fast and small DNA sequencing

1. Double- or single-stranded DNA is sequenced using the dideoxy chain termination method (27).

2. Prepare the solution for pouring the gel by mixing 13.5 ml of acrylamide solution (6.66% acrylamide, 0.35% bisacrylamide, 48% urea), 1.5 ml of 10× Tris–borate buffer (12.1% Tris, 6.2% boric acid, 0.75% EDTA), 12.5 μl TEMED (*N,N,N′,N′*-tetramethylethylenediamine), and 57.5 μl of 15% ammonium persulfate.

3. Cast and pour the gel. We used a shark teeth comb that was cut to fit the size of the gel. After pouring the gel, the comb was inserted upside down until polymerization.

4. Place the gel in the electrophoresis apparatus and pre-run at 1000 V for 5–10 min.

5. Load the samples and run at 1000 V. In these conditions the bromophenol blue should reach the front in about 20 min and the xylene cyanol in 40 min.

6. After electrophoresis remove the gel and do not fix. Dry and expose the gel. Obviously the smaller size of the gel requires a shorter drying time.

---

6. Comparison of helical stability of DNA fragments

Gradient denaturing gel electrophoresis is a useful technique to compare the helical stability of DNA fragments (28). When DNA is electrophoresed in a polyacrylamide gel in the presence of a gradient of urea and formamide just below its melting temperature, it moves as a helical duplex until it encounters the concentration of denaturants that results in its melting. At this point the DNA has portions that are helical duplexes and others that are random coils. As a consequence, the speed of migration is significantly reduced. Therefore two DNA fragments of the same size but with different helical stabilities will be resolved, with the less stable fragment being the slower band when electrophoresed under these conditions. This technique was applied to evaluate alterations in DNA helix stability due to base modifications and to detect and locate single base changes in a stretch of DNA (28, 29). Recently Gammie and Crosa (30) applied this method to compare the helical stability of the methylated versus unmethylated origin of replication DNA region from the plasmid pColV-K30. The result of that experiment is shown in *Figure 3*. When run in a parallel gradient (30–60%) of denaturants, methylated DNA was retarded as compared to the same unmethylated DNA fragment, demonstrating that methylation contributes significantly to the helical instability of the origin of replication region of pColV-K30. A thorough description of this method has been recently published (29).

You will need the following equipment:

- a set-up for polyacrylamide gel electrophoresis with a circulating water temperature control system (29)
- a gradient maker
- a power supply that can provide 150 V

6.1 Gel preparation and electrophoresis

The reagents to be used are:

- 20 × TAE electrophoresis buffer (0.8 M Tris base, 0.4 M sodium acetate, 20 mM EDTA pH 7.4)

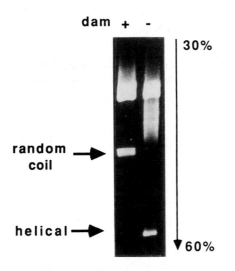

Figure 3. Helical stability of methylated and unmethylated pColV-K30 origin DNA. Methylated (+) and unmethylated (−) DNA from a recombinant plasmid generated by cloning a pColV-K30 DNA fragment including the origin of replication in the pBluescript vector was digested with *Xba*I and *Hind*III to release a 500 bp portion containing the minimum origin of replication (*lower bands*) and run through a polyacrylamide gel with a parallel gradient of denaturants ranging from 30 to 60%. [Reproduced from Gammie, A. E. and Crosa, J. H. (1991). *Molec. Microbiol.*, **5**, 495, with permission.]

- acrylamide stock solution (40% acrylamide) (37.5:1 acrylamide:bisacrylamide)

- denaturing stock solutions (0% denaturant: 6.5% acrylamide in TAE; 100% denaturant: 6.5% acrylamide, 7 M urea, 40% formamide)—denaturing stock solutions are made by adding 16.2 ml acrylamide stock, 5 ml 20 × TAE together with urea and formamide (for 80% denaturing solution add 33.6 g urea and 32 ml formamide) and make up to 100 ml with distilled water.

- 20% ammonium persulfate freshly prepared

- TEMED

Protocol 15. Comparison of helical stability of DNA fragments

1. Prepare the parallel gradient gel by starting with a gel containing a gradient from 0% to 80% denaturing agent, and according to the results expand by making a new gradient gel using a gradient in the appropriate range, for example 30% to 60%. For a 25 ml gel volume and a gradient 0% to 80% prepare 12.5 ml of 0% and 80% denaturing solutions, add 62.5 µl of 20% ammonium persulfate, and 6 µl TEMED.

2. Pour the solutions in the appropriate chambers of the gradient maker, cast the gel with the help of a peristaltic pump, and allow the gel to polymerize.

3. Set up the apparatus, adjust the temperature to 60°C, load the samples, and apply 100–150 V. The time of the run depends on the size of the DNA fragment.

4. Stain with ethidium bromide to visualize the plasmid DNA bands.

---

7. Identification of plasmid-encoded products

The proteins and/or RNAs, can be identified and characterized by using different experimental approaches. Plasmid-encoded proteins are usually identified by using *in vitro* or *in vivo* systems, such as prokaryotic DNA-directed translation and maxicells or minicells methods, respectively. Plasmid-encoded RNAs can be characterized by northern and RNase protection analysis. We describe here the protocols we used to characterize the products encoded by the virulence plasmid pJM1.

Protocol 16. Immunoprecipitation of *in vitro* coupled transcription–translation analysis of plasmid-encoded proteins

We will not describe the preparation and the reaction conditions of this system since commercial kits (Prokaryotic DNA-Directed Translation Kit, Amersham) are already available. Described here are the conditions used to identify and isolate a particular polypeptide encoded by a plasmid from the transcription–translation reaction mixture. The equipment required for this analysis is that described in Section 5.

After the plasmid-directed transcription–translation reaction was completed and the incorporation of the radioactive amino acid precursor was determined, all the radiolabelled polypeptides synthesized *in vitro* can be analysed by SDS polyacrylamide gel electrophoresis (SDS–PAGE) and subsequent detection by fluorography (31). The synthesis of a particular protein can be assayed by immunoprecipitation from the transcription–translation mixture, with a specific antiserum, as follows.

1. Transfer 35 μl of the transcription–translation mixture to a micro-centrifuge tube and add 1.5 ml of NIBBT solution (50 mM Tris–HCl pH 7.4, 150 mM NaCl, 5 mM EDTA, 1.0% Triton X-100), and 100 μl of a 10% suspension of heat-killed, formalin-fixed *Staphylococcus aureus* Cowan strain I cells (32). Incubate the suspension at room temperature for 30 min and centrifuge at 12 000 r.p.m. for 20 min at 4°C.

2. Transfer the supernatant to a fresh tube, add 1/500 volume of specific antiserum, and then incubate the reaction mixture overnight at room temperature.

3. Add 0.15 ml of the 10% suspension of fixed *Staphylococcus aureus* Cowan strain I, and continue the incubation at room temperature for 30 min.

4. Recover the fixed bacterial cells carrying the immunocomplexes by centrifugation as described in step **1**, and wash the pellet four times with NIBBT.

5. Resuspend the pellet in 100 μl of SDS–PAGE sample buffer and incubate in a boiling water-bath for 5 min. Load and electrophorese the sample, detecting the radioactive protein bands by fluorography. Controls should include the transcription–translation of the cell-free extract containing either no plasmid or the plasmid vector used for gene cloning. The molecular masses of the radioactive proteins can be calculated using as reference radiolabelled molecular weight standard proteins. Alternatively, non-labelled standard proteins can be used, but in this case the gel should be stained before it is processed for fluorography.

Protocol 17. Maxicells analysis

1. Prepare a 5 ml L broth overnight culture at 37°C using one colony of the *E. coli* maxicell strain BN660 (33) harbouring the plasmid to be analysed.

2. Using a 125 ml flask, inoculate 5 ml of fresh 'K' medium (M9 minimal medium (18) containing 4 mg/ml thiamine and 1% casamino acids) with 100 μl of the overnight culture, and incubate at 37°C with good shaking until an OD_{600} value of 0.5 is reached.

3. Transfer 100 μl of this culture to 10 ml of K medium in a 250 ml flask and incubate as described in the previous step.

4. Transfer the culture to a Petri dish, remove the cover, and irradiate the cells for 2 min using an UV lamp (254 nm). For this purpose use a rotator plate, rotating the culture gently during the irradiation. Locate the plate 15 cm from the UV lamp, and perform this step in a dark-room.

5. Transfer aseptically the culture to a culture tube previously covered with aluminium foil and incubate in a shaker at 37°C for 2 h.

6. Add 100 μl of a freshly prepared solution of 25 mg/ml cycloserine and continue the incubation overnight.

7. Add another 100 μl of a freshly prepared solution of 25 mg/ml cycloserine and incubate the cells for an additional 60 min at 37°C.

8. Transfer 1.5 ml of culture to a microcentrifuge tube and collect the cells by centrifugation at 6000 r.p.m. at 4°C. Wash the cells twice with 1 ml of Hershey salts (5.4 g NaCl, 3.0 g KCl, 1.1 g NH_4Cl, 15 mg $CaCl_2.2H_2O$, 0.2 g $MgCl_2.6H_2O$, 0.2 mg $FeCl_3.6H_2O$, 12.1 g Tris base, per litre, pH 7.4).

9. Resuspend the cells in 750 μl of Hershey medium (Hershey salts containing 0.4% glucose, 0.01% threonine and leucine, and 0.02% proline and arginine), transfer to a 10 ml plastic tube, and incubate for 1 h at 37°C.

10. Add 25 μCi of [^{35}S] methionine and incubate for 30 min at 37°C.

11. Transfer the cells to a microcentrifuge tube, centrifuge as described in step **8**, washing the cells with 1 ml of Hershey salt.

12. Resuspend the pellet in 40 μl of sample buffer and dissolve the cells by incubation in a boiling water-bath for 5 min.

13. Analyse the plasmid encoded proteins by SDS–PAGE, loading 5–10 μl of each radiolabelled sample. Control samples should include the maxicell strain harbouring either no plasmid or the cloning vector. The size of the labelled proteins can be determined as described for the *in vitro* coupled transcription–translation protocol.

8. RNA extraction and analysis

Precautions must be taken when working with RNA to avoid the action of RNases (18). Glassware must be baked at 180°C for at least 8 h or filled with 0.1% diethyl pyrocarbonate (DEPC), let stand 2 h, rinsed several times with water, and autoclaved for 1 h (liquid cycle). The sterile, disposable plastic-ware is normally free of RNases and can be used without prior treatment. An RNase-free electrophoresis tank should be reserved exclusively for RNA work. Solutions should be made with DEPC treated water (DEPC is added to water to make a 0.1% solution, let stand 12 h, followed by autoclaving for 1 h on liquid cycle). Bottles of reagents should be opened and kept exclusively for solutions to work with RNA. All RNA work should be performed with gloves.

8.1 RNA extraction

Protocol 18. RNA extraction (34)

1. Grow cells in 5 ml of media with the appropriate antibiotic overnight and use this culture to inoculate 400 ml of media. Grow bacteria until the OD_{600} is approximately 0.4.

2. Spin down the cells at 3000 g for 5 min. Pour off the supernatant. Starting at this point it is convenient to keep the samples either at 65°C or 4°C when possible.

3. Pre-heat a tube with 4 ml of phenol (phenol for RNA extraction consists of phenol equilibrated with 0.1 M sodium acetate pH 4.5) and the lysis solution (0.15 M sucrose, 10 mM sodium acetate pH 4.5, 1% sodium dodecyl sulfate) at 65°C.

4. Add 4 ml lysis solution to the cells using a plastic pipette and re-suspend by pipetting up and down repeatedly. Transfer immediately to the tube containing the heated phenol.

5. Mix gently and keep the tube in the 65°C bath. Mix several times during a period of 5 to 10 min, and centrifuge at 27 000 g for 5 min at 4°C.

6. Transfer the aqueous (upper) phase to another tube containing hot phenol and repeat step **5**.

7. Transfer the upper phase (do not take the white protein layer) to a tube containing phenol/chloroform/isoamyl alcohol (50:48:2). Repeat step **5**.

8. Transfer upper phase to a tube containing chloroform/isoamyl alcohol (24:1). Repeat step **5**.

9. Remove the chloroform layer from below the sample.

Protocol 18. *Continued*

10. Add 0.5 volumes of 6 M ammonium acetate (should be about 1.2 ml) and 2–3 volumes of 95% ethanol. Keep the tubes at −20°C for 15 min.

11. Spin down the RNA at 17 000 r.p.m. for 10 min and wash twice the RNA pellet with 70% ethanol. Dry the pellet.

12. Resuspend in 500 μl of water, transfer 250 μl to two microcentrifuge tubes. Add 250 μl of phenol/chloroform/isoamyl alcohol (50:48:2) to each tube. Mix gently and centrifuge in a microcentrifuge.

13. Transfer the aqueous phase to another tube and add one volume of chloroform/isoamyl alcohol (24:1). Centrifuge in a microcentrifuge.

14. Transfer 200 μl to another tube, add 100 μl of 6 M ammonium acetate, and 900 μl of 95% ethanol. Centrifuge in the microcentrifuge, and wash the pellet twice with 70% ethanol. Dry and resuspend the pellet in 1 ml of water.

15. Separate in 100 μl aliquots and keep at −70°C. Use one aliquot to determine quantity and quality of the RNA.

16. Measure OD_{258}. Calculate concentration RNA ($\mu g/ml = OD_{258} \times$ dilution factor $\times$ 0.04).

17. Measure OD_{260} and determine the ratio OD_{260}/OD_{258}. The value should be between 1.7 and 2. In the case of a value smaller than 1.7 another phenol extraction is recommended.

8.2 Northern blot hybridization

The RNA is separated using electrophoresis in denaturing agarose gels. The denaturant can be either formaldehyde or glyoxal/dimethylsulfoxide. Here we describe the protocol using formaldehyde we normally use (18, 35). After electrophoresis, the RNA is transferred to a nitrocellulose or nylon membrane and hybridized to a specific DNA or RNA labelled probe.

Protocol 19. Electrophoresis in agarose/formaldehyde gel and northern blot hybridization

1. Prepare a 1.2% agarose solution in [3-(*N*-morpholino)-propanesulfonic acid] (MOPS) buffer (10 × MOPS solution is prepared by adding 41.8 g of MOPS to 800 ml of water, adjusting the pH to 7.0, followed by addition of 16.6 ml of 3 M sodium acetate and 20 ml of 0.5 M EDTA pH 8.0, and bringing the solution to 1 litre with water) by adding 30.4 ml of water, 0.42 g of agarose, 3.5 ml of 10 × MOPS. Melt the agarose and add 1.1 ml of 37% formaldehyde. Pour the gel. Running buffer is made

by mixing 50 ml 10 × MOPS, 42.5 ml 37% formaldehyde, and bringing the solution to 0.5 litre.

2. Prepare and load the samples. Samples are prepared by adding 16 μl of sample buffer (5 ml 10 × MOPS, 8.7 ml of 37% formaldehyde, 5 ml formamide) to 10 μl of RNA solution (10–20 μg), heating at 55°C for 15 min, followed by addition of 4 μl of loading buffer (50% glycerol, 1 mM EDTA pH 8.0, 0.25% bromophenol blue). It is convenient to run duplicates to check the RNA after electrophoresis.

3. Electrophorese at about 5 V/cm. Stain one of the duplicates and blot the other. To stain rinse the gel several times with water and submerge in an ethidium bromide solution (25 μl of 5 mg/ml ethidium bromide in 500 ml of water). Let the gel stain in the dark and photograph the RNA on top of a UV box. It may be necessary to destain the gel by placing it in water for 1 h.

4. To transfer the RNA first rinse the gel with 10 × SSC (to make 20 × SSC dissolve 87.6 g of sodium chloride, and 44 g of sodium citrate in 400 ml of water, adjust pH to 7.0 with either NaOH or HCl, and bring the solution to 500 ml with water).

5. Blot the gel by preparing a set-up as shown in *Figure 4* and leaving it overnight. The blotting buffer is 10 × SSC.

6. Take the membrane and bake for 2 h at 80°C or treat under a UV cross-linker device.

7. Place the membrane in a plastic bag, add 10 ml of pre-hybridization solution (25 mM phosphate buffer pH 7.4, 5 × SSC, 5 × Denhardt's solution [50 × Denhardt's stock solution: 1% Ficoll, 1% polyvinyl-pyrrolidone, 1% bovine serum albumin Pentax fraction V, keep at −20°C], 50 μg/ml salmon sperm DNA), seal the bag, and let stand for 2 h at a temperature between 37°C and 42°C.

8. Pour off pre-hybridization solution, add 10 ml hybridization solution (pre-hybridization solution with the addition of 10% dextran sulfate), and add the denatured labelled probe. Let stand overnight at the same temperature as in the pre-hybridization.

9. Take the membrane and wash once with 1 × SSC/0.5% sodium dodecyl sulfate for 5 min at room temperature, twice with 0.1 × SSC/0.1% sodium dodecyl sulfate/5 mM EDTA at 67°C for 1 h, and twice with 0.1 × SSC at 67°C for 1 h. Let the membrane dry and expose to X-ray film.

8.3 Primer extension

This technique is applied to identify the 5′ end of an RNA species and consists of the incubation of a labelled single-stranded DNA primer to the extracted

18. Sambrook, J., Fritsch, E., and Maniatis, T. (ed.) (1989). *Molecular cloning, a laboratory manual*. Cold Spring Harbor Press, Cold Spring Harbor, NY.
19. Rose, M. D. and Fink, G. R. (1987). *Cell*, **48**, 1047.
20. Gammie, A., Tolmasky, M. E., and Crosa, J. H. (1993). *J. Bacteriol.*, **175**, 3563.
21. Prentki, P. and Krish, M. (1984). *Gene*, **29**, 303.
22. Actis, L. A., Tolmasky, M. E., Farrel, D. H., and Crosa, J. H. (1988). *J. Biol. Chem.*, **263**, 2853.
23. Ruvkun, G. and Ausubel, F. (1981). *Nature*, **289**, 85.
24. Tolmasky, M. E., Salinas, P., Actis, L. A., and Crosa, J. H. (1988). *Infect. Immun.*, **56**, 1608.
25. Knauff, V. and Nester, E. (1982). *Plasmid*, **8**, 45.
26. Hirsch, P. R. and Beringer, J. E. (1984). *Plasmid*, **12**, 139.
27. Sanger, F., Nicklen, S., and Coulson, A. R. (1977). *Proc. Natl. Acad. Sci. USA*, **74**, 5463.
28. Collins, M. and Myers, R. (1987). *J. Mol. Biol.*, **198**, 737.
29. Myers, R., Maniatis, T., and Lerman, L. (1987). In *Methods in enzymology* (ed. R. Wu), Vol. 155, pp. 501–27. Academic Press London.
30. Gammie, A. and Crosa, J. H. (1991). *Mol. Microbiol.*, **5**, 495.
31. Actis, L. A., Tolmasky, M. E., Farrel, D. H., and Crosa, J. H. (1988). *J. Biol. Chem.*, **263**, 2853.
32. Kessler, S. W. (1975). *J. Immunol.*, **115**, 1617.
33. Sancar, A., Hack, A. M., and Rupp, W. D. (1979). *J. Bacteriol.*, **137**, 692.
34. von Gabian, A., Belasco, J., Schottel, J., Chang, A., and Cohen, S. (1986). *Proc. Natl. Acad. Sci. USA*, **80**, 653.
35. Waldbeser, L., Tolmasky, M., Actis, L., and Crosa, J. (1993). *J. Biol. Chem.*, **268**, 10433.
36. Ghosh, P., Reddy, V., Swinscoe, J., Lebowitz, P., and Weissman, S. (1978). *J. Mol. Biol.*, **126**, 813.

5

Plasmids in *Agrobacterium tumefaciens* and *Rhodococcus fascians*

JAN DESOMER, ROLF DEBLAERE, and
MARC VAN MONTAGU

1. Introduction

In this chapter, we would like to provide an overview of the main techniques used in our laboratory to identify, isolate, and analyse plasmids (and the functions they encode) in plant-tumourigenic bacteria from the Gram-negative genus *Agrobacterium* and from the Gram-positive genus *Rhodococcus*. Although these bacteria are evolutionary far apart, in both cases phytopathogenicity is determined by large conjugative plasmids, and similar genes have been detected on plasmids from both groups of bacteria (1). The plasmid-based recombinant DNA technology developed for *Agrobacterium* can also be applied to other bacteria that interact with plants, such as the nitrogen-fixing rhizobacteria (2).

1.1 *Agrobacterium tumefaciens*

A. tumefaciens and *A. rhizogenes* are soil phytopathogens that can genetically transform wounded plant cells, and thus provide ideal vector systems for the generation of transgenic plants. An overview of all the new transgenic plant species and their agro-industrial applications is beyond the scope of this Chapter and can be found in several reviews (3, 4).

Genetic and molecular analysis of the *Agrobacterium*/plant interaction led to the definition of precise segments of the tumour-inducing (Ti) or root-inducing (Ri) plasmids that are transferred to the plant cell. Integration in the nuclear genome and expression of this transferred DNA (T-DNA) results in the crown gall phenotype. This interaction is coordinated by a biochemical communication process between susceptible plant cells and the infecting bacteria that is mediated by the gene products from the virulence (*vir*) region on the Ti plasmid. Sensing of the plant cell-derived signal molecules by the bacteria leads to the excision of the T-DNA from the Ti plasmid, followed by subsequent transfer and integration in the infected plant cell genome (for review, see 5). Exploitation of this knowledge resulted in the design of modified, non-tumourigenic *Agrobacterium* Ti plasmids for the engineering of transgenic plants.

Ti plasmids as well as Ri plasmids are classified according to the T-DNA-directed production by the transformed plant cell of specific metabolites (opines) which are metabolized by the infecting bacteria (6). The T-DNAs of the two most studied types of Ti plasmids (octopine- and nopaline-type Ti plasmids) have a central DNA segment in common as well as a 25 bp border sequence. This central region encodes the proteins responsible for the bio-synthesis of the phytohormones auxin and cytokinin, which are responsible for the crown gall phenotype (7, 8). However, inactivation or deletion of these plant 'oncogenes' does not affect the transfer and integration of the mutant T-DNA into the plant cell genome, but it modifies or abolishes the disease symptoms (9). This crucial finding, together with the development of selectable markers for transformed plant cells, enabled the *Agrobacterium* system to be successfully used for gene transfer to plants.

A wide variety of non-oncogenic *Agrobacterium* vector systems has been developed. In the cointegration-type vectors, the T-DNA genes were re-placed by the *Escherichia coli* cloning vector pBR322. Foreign or chimeric genes, cloned into pBR-derived intermediate vectors, can be introduced into *Agrobacterium* and allowed to cointegrate in the resident, non-oncogenic Ti plasmid. A more detailed overview of these vectors is presented by Deblaere *et al.* (10).

The development of binary vectors is based on the observation that the T-DNA can be physically separated from the Ti plasmid, for example by cloning on a separate replicon, without affecting transfer to the plant cell (11). This is now the most frequently used vector system. It consists of a cloning vehicle capable of replicating in *E. coli* and *Agrobacterium* and contains unique cloning sites and plant selectable marker(s) between the T-DNA border sequences (a typical example is presented in *Figure 1*). The second element is a Ti plasmid from which the T-DNA has been deleted, such as pGV2260 (12). It provides the *vir* functions necessary for T-DNA transfer. Commonly used binary vectors are derived from Inc P-type plasmids (e.g. RP4) that have a broad host range. However, due to segregational and/or structural instability of these plasmids in *Agrobacterium*, a new set of binary T-DNA vectors, based on the replication origin of the *Pseudomonas* plasmid pVS1 was de-veloped. These plasmids are very stable in *Agrobacterium* cultures grown under non-selective conditions. For replication in *E. coli* the pBR322-derived origin was included (see *Figure 1*).

Foreign DNA or a gene of interest is cloned into the binary vector between the T-DNA borders and the recombinant plasmid is introduced into *Agro-bacterium* via genetic or *in vitro* techniques (see Section 4.1). Because co-integration in a resident Ti plasmid is no longer required, transconjugants or

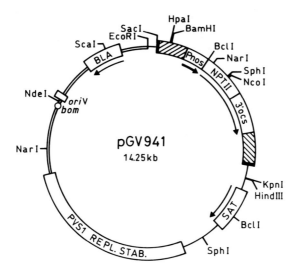

Figure 1. Prototype of a pVS1-based binary T-DNA vector. The construction of pGV941 has been described elsewhere (10). The *fine double line* indicates pBR322 sequences containing the origin of replication (*oriV*) and the basis of mobilization site (*bom*). The region encoding the replication and stability functions derived from the pVS1 plasmid is indicated. The *hatched region* represents T-DNA-derived sequences containing the 25 bp border sequences (*black boxes*). Abbreviations: Pnos, nopaline synthase promoter; NPTII, promoterless *E. coli* neomycin phosphotransferase II gene; 3' ocs, octopine synthase polyadenylation region; BLA, β-lactamase gene; SAT, streptomycin acetyltransferase gene.

transformants can be selected with 100 to 1000-fold higher frequencies compared to that of cointegration vectors. Consequently, *Agrobacterium* can be used as a direct host for the construction of a genomic plant DNA library between T-DNA borders.

1.2 *Rhodococcus fascians*

Rhodococcus fascians is a Gram-positive nocardioform phytopathogen (13) that causes fasciation—the loss of apical dominance and the development of abundant small shoots—upon infection of a wide range of dicotyledonous and some monocotyledonous plants (14–16). In addition to this plant pathogen, the genus *Rhodococcus* contains some species that are pathogenic for animals and humans as well as a wide range of bacteria with specific metabolic traits of potential industrial interest (for a review, see 17).

Two types of plasmid have been identified in *R. fascians*. Several strains contain large covalently closed circular (CCC) plasmids (18–20; see also *Table 1*). Plasmids such as pRF2 or pD188 are self-transmissible between strains of *R. fascians* and encode chloramphenicol and/or cadmium resistance (20). The second type of extrachromosomal element in *R. fascians* is unique to virulent strains. These plasmids are large (~ 200 kb), conjugative, linear DNA molecules (1). It has not yet been established whether they belong to the class of linear plasmids found in *Borrelia* (possessing true telomeres) or to the invertron-type of linear plasmids that have proteins covalently attached to their 5′ ends and which are found in a number of prokaryotes (*Streptomyces* spp.) and eukaryotes (fungi, plant mitochondria) (reviewed in 21).

It is clear however that these linear plasmids encode essential fasciation genes. Insertion mutagenesis of one of these fasciation-inducing (Fi) plasmids revealed that at least three loci (*fas, att, hyp*) are involved in phytopathogenicity (1). It has been demonstrated that *fas* codes for an isopentenyltransferase (the first step in cytokinin biosynthesis), but less is known about the other loci (1).

2. Identification of endogenous plasmids by small-scale isolation procedures

2.1 *Agrobacterium tumefaciens*

A common characteristic of bacteria belonging to genera such as *Agrobacterium* or *Rhizobium* is the presence of very large plasmids. For example, tumourigenic *A. tumefaciens* strains contain Ti plasmids with an average size of ~ 200 kb. Several procedures have been described for the isolation of such large plasmids. This section will focus on two methods which were developed for analytical purposes (see *Protocols 1* and *2*). The last method describes a microscale isolation procedure for small plasmids such as commonly used binary vectors.

2.1.1 Eckhardt-type analysis

The first technique, also referred to as Eckhardt-type gel analysis, allows the determination of the number of different plasmids and an estimation of their respective sizes in a given strain. In essence, lysozyme treated cells are lysed directly in the wells of an agarose gel by the action of sodium dodecyl sulfate

Table 1. Large plasmids in *R. fascians* (20)

Strain	Plasmid	Size (kb)
D188	pD188	138
NCPPB 1488	pD188	138
PD299	pRF1	73
PD300	pRF1	73
PD301	pRF1	73
NCPPB 1675	pRF2	160
NCPPB 2551	pRF3	140
NCPPB 469	pRF4	100

under the influence of an electrical field and the released plasmids are then separated by electrophoresis. In the original procedure a vertical gel electrophoresis set is used. Here we describe a variant of this method which relies on the use of a horizontal submerged-type of agarose gel electrophoresis (*Protocol 1*).

Protocol 1. Identification of Ti plasmids by Eckhardt-type analysis (modification of the protocol described by Eckhardt [22])

- medium size gel tray (15 cm×15 cm), two well combs, submerged gel electrophoresis apparatus
- agarose (0.65%) dissolved in TBE buffer (89 mM Tris base, 89 mM boric acid, 2.5 mM EDTA, pH 8.2)
- solution I: 25% sucrose, 1 U/μl RNase in TBE buffer—boil for 10 min, dispense in aliquots, and store at −20°C: just before

- use, add lysozyme to a final concentration of 100 μg/ml
- solution II: 1% agarose in TBE buffer supplemented with 0.1% bromophenol blue, 2% SDS in TBE buffer—mix equal portions of these two solutions and dispense in aliquots: boil just before use
- ethidium bromide (0.5 μg/ml)

A. *Gel preparation*

1. Pour agarose solution into the tray to a thickness of ∼ 7 mm.
2. Insert two combs, back to back. The cathode proximal comb should be inserted 2 mm deeper than the cathode distal one.
3. When solidified, remove the cathode proximal comb and fill the wells with solution II.
4. When set, remove the second comb and insert the gel in a submerged electrophoresis tank. About 2–3 mm of TBE buffer should cover the gel.

B. *Sample preparation*

1. Grow a 5 ml cell culture to mid-log or early stationary phase.
2. Standardize cell density to 3 × 10^8 cells/ml.
3. Centrifuge 1 ml of cells in an Eppendorf tube for 2 min at 13 000 r.p.m. Discard the supernatant, briefly spin the tubes again, and remove all remaining liquid.
4. Resuspend the bacterial pellet in 20 μl of solution I.

C. *Sample application and electrophoresis*

1. Load the resuspended bacteria *directly* into the wells.
2. Immediately apply the current: apply 5–10 mA for ∼ 30 min, until the bacteria-containing wells are cleared. Then increase the current to ∼ 50 mA and continue for 3 h.

method is very reproducible, extremely fast, and tends to yield higher amounts of plasmid DNA than *Protocol 2*.

Table 2. Antibacterial drugs for *Agrobacterium*

Antibiotic	Final concentration (μ/ml)	Solvent	Comments
Tricarcillin	100	H_2O	Best results in YEB medium.
Chloramphenicol	25–50	Ethanol	Poor marker; use in combination with another antibiotic.
Erythromycin	50	Ethanol	
Gentamycin	40	H_2O	
Kanamycin	25	H_2O	Not usable in minimal medium.
Neomycin	40	H_2O	500 μg/ml in minimal medium.
Rifampicin	100	DMSO	
Spectinomycin	100	H_2O	300 μg/ml in minimal medium.
Streptomycin	300	H_2O	1000 μg/ml in minimal medium.
Sulfonamide	250	DMSO	To be used in minimal or Müller–Hinton medium.
Tetracycline	2.5	DMSO	Poor marker.

2.2 *Rhodococcus fascians*

2.2.1 Linear plasmids

Large linear plasmids can be visualized routinely by pulsed-field gel electrophoresis (PFGE). These electrophoresis methods in which electric fields are inverted (field inversion gel electrophoresis, FIGE) or rotate around the agarose gel (countour-clamped electric field, CHEF), allow the separation of large linear DNA molecules, while large CCC plasmids do not migrate in the gel. However, small CCC plasmids do migrate, but they can be recognized by modifying the electrophoresis conditions (pulse time, ratio forward versus inverse electrical field, angle of rotation of electrical field). Under these modified conditions small CCC plasmids move at different positions relative to linear marker DNAs (λ multimers or *Saccharomyces cerevisiae* chromosomes).

Since intact *R. fascians* cells are recalcitrant to the action of ionic detergents, such as SDS, bacterial cells have to be pre-treated with lysozyme and polyethylene glycol (PEG 6000), according to a method originally described by Chassy (23) for oral streptococci (see *Protocol 3*). This treatment yields detergent-sensitive bacteria, which remain, however, resistant to osmotic shock (20).

Protocol 3. Sensitization of *R. fascians* cells to detergent action

- TRE buffer: 50 mM Tris–HCl pH 8.0, 20 mM EDTA
- lysozyme
- polyethylene glycol (PEG 6000) (25%) solution in TRE buffer

Method

1. Grow *R. fascians* in YEB medium (see *Table 4*) at 28°C on an orbital shaker (200 r.p.m.) to an OD_{600} of 0.7–0.8 (typically two days when starting from a loopful inoculum).

2. Harvest cells by centrifugation and wash in TRE buffer.

3. Resuspend bacterial cells in 1/10 of the original volume of TRE to which 1 mg/ml lysozyme was added.

4. Add four volumes (4/10 of the original volume) of 25% PEG 6000 dissolved in TRE buffer.

5. Incubate at 37°C for at least 2 h.

6. Harvest cells by centrifugation, wash with TRE, and resuspend at the desired density in TRE or appropriate solution.

Remarks

(a) This protocol equally applies to other rhodococci (J. Desomer, unpublished data).

(b) Prior to step **6,** an aliquot of the mixture can be tested for detergent sensitivity, allowing further incubation with additional lysozyme, if necessary.

Care has to be taken to avoid shearing of large linear DNA molecules, so lysis is accomplished after immobilizing detergent-sensitive *R. fascians* cells in low temperature-gelling agarose plugs (see *Protocol 4*).

Protocol 4. Lysis of *R. fascians* cells in agarose plugs (1)

- low-gelling agarose (0.7%) in 125 mM EDTA
- ES buffer: 125 mM EDTA, 1% Sarkosyl
- proteinase K

Method

1. Grow *R. fascians* at 28°C in YEB medium to an OD_{600} of 0.7–0.8.

2. Harvest the cells from a 5 ml culture by centrifugation and treat them according to *Protocol 3* to obtain detergent-sensitive cells.

Protocol 7. Small scale isolation of CCC plasmids in
R. fascians

- TRE buffer (see *Protocol 3*)
- solutions B and C (see *Protocol 2*)
- isopropanol
- Tris–HCl saturated phenol
- chloroform
- ethanol (70%)
- RNase solution (1 mg/ml)

Method

1. Grow *R. fascians* cells in 50 ml YEB at 28°C (two days).

2. Harvest cells by centrifugation in a fixed-angle rotor (7000 *g*).

3. Prepare detergent-sensitive cells as described in *Protocol 3*.

4. Resuspend cells in 3 ml of TRE buffer.

5. Add 3 ml of solution B. Allow to stand at room temperature for 10 min.

6. Add 3 ml of solution C. Mix gently by inversion and incubate on ice for 1 h.

7. Centrifuge at 13 000 *g* in a fixed-angle rotor in a Sorvall-type centrifuge.

8. Collect cleared supernatant in a plastic centrifuge tube (30 ml) and precipitate nucleic acids by addition of 0.7 volume isopropanol.

9. Pellet nucleic acids by centrifugation (7000 *g*).

10. Briefly dry pellet in a vacuum desiccator and resuspend in 1 ml of TRE buffer.

11. Add 5 μl of a 1 mg/ml RNase solution and incubate for 10 min at 37°C.

12. Extract the DNA solution with phenol once, phenol/chloroform once, and chloroform. Precipitate DNA with 0.7 volume isopropanol.

13. Pellet the DNA by centrifugation in a microcentrifuge and wash the pellet with 70% ethanol.

14. Dry the pellet and resuspend in 50 μl of H_2O.

The plasmid obtained can be visualized using conventional agarose gel electrophoresis, after restriction enzyme digestion if appropriate. Half of the preparation should be used for one analysis. The DNA can also be used to electrotransform *R. fascians* after dialysis (see *Protocol 14*).

3. Large scale purification methods

3.1 Large scale Ti plasmid preparation from *Agrobacterium*

For a detailed physical analysis of the Ti plasmid such as extensive restriction enzyme mapping, preparation of radioactively labelled probes, or for sub-cloning experiments, large amounts of highly purified plasmid DNA are required. The protocol presented below is essentially as described by Hirsh *et al.* (25) and was further developed in our laboratory.

Protocol 8. Isolation of Ti plasmids

- TE buffer (see *Protocol 2*)
- Pronase solution (5 mg/ml in TE buffer, pre-digested for 1 h at 37°C)
- SDS (10%) in TE buffer
- NaOH (3 M)
- Tris–HCl (2 M) pH 7.0

- NaCl (5 M)
- PEG 6000 (50%)
- CsCl
- ethidium bromide
- TE saturated isobutanol
- ethanol

Method

1. Inoculate a 10 ml *Agrobacterium* culture in liquid PA medium and incubate overnight at 28°C and shaking at 250 r.p.m.

2. Inoculate one litre of PA medium (in a 4 litre Erlenmeyer) with the 10 ml saturated pre-culture.

3. Incubate the culture at 28°C on an orbital shaker (250 r.p.m.) for 24 h. Good aeration is essential.

4. Harvest the bacteria, (e.g. a GSA rotor in a Sorvall centrifuge) by spinning 10 min at 2000 *g*.

5. Wash the cells in 200 ml TE buffer.

6. Centrifuge and resuspend the cells in 80 ml TE buffer.

7. Add 10 ml Pronase solution and 10 ml 10% SDS.

8. Incubate the lysate for 1 h at 37°C. If necessary, continue until the lysate is clear.

9. Transfer the lysate to a sterile 250 ml beaker.

10. Adjust the clear, viscous lysate to pH 12.4 by slowly adding 3 M NaOH with gentle stirring.

11. Leave for 30 min at room temperature with slow stirring.

12. Adjust the lysate to pH 8.5 with 2 M Tris–HCl pH 7.0. The lysate

---

Protocol 8. *Continued*

should become non-viscous (if still viscous, the denaturation—neutralization step should be repeated).

13. Transfer the lysate to a 250 ml centrifuge tube.

14. Add 5 M NaCl to a final concentration of 1 M.

15. Mix by *gentle* inversion.

16. Leave on ice for at least 4 h (or overnight).

17. Precipitate the SDS/NaCl complex (chromosomal DNA, proteins, and cell debris) by centrifugation for 20 min at 16 000 g and 4 °C.

18. Decant the supernatant into a fresh tube.

19. Add 50% PEG 6000 (w/v) to a final concentration of 10% and mix gently.

20. Leave on ice overnight.

21. Pellet the DNA by centrifugation (GSA rotor: 10 min at 8000 g and 4 °C).

22. Discard supernatant and dissolve the DNA pellet in 2.5 ml TE. Do this extremely carefully, because large DNA molecules are very sensitive to shearing.

23. Add 6.1 ml TE buffer, 9.3 g CsCl, and 0.5 ml ethidium bromide (10 mg/ml).

24. Mix by gentle inversion.

25. Transfer to a Beckman Quick-Seal[RM] tube (16 mm × 76 mm).

26. Centrifuge in a Beckman rotor NVT65 for 24 h at 48 000 r.p.m. and 15 °C.

27. Illuminate the gradient under UV light.

28. Remove the lower band containing the supercoiled DNA from the side of the tube with a 1 ml syringe, using a wide needle. First puncture the top of the tube as an air inlet.

29. Remove ethidium bromide by five extractions with TE saturated iso-butanol.

30. Transfer the DNA sample to a 15 ml Corex tube and add two volumes of TE buffer.

31. Add ethanol equal to two times the volume after dilution, and precipitate the DNA overnight at −20 °C.

32. Centrifuge the ethanol precipitate in a Sorvall HB4 rotor for 20 min at 13 000 g.

33. Dry the DNA pellet and resuspend gently in 0.5 ml TE buffer.

34. Precipitate again in an Eppendorf tube by adding two volumes of ethanol.

35. Dissolve the pellet at an approximate concentration of 100 μg/ml. 50 μg of plasmid DNA can be obtained starting from a 1 litre culture.

Remark
Before starting the CsCl gradient ultracentrifugation, 50 μl of the crude plasmid preparation can be analysed by electrophoresis on a 0.6% agarose gel. This will show the content of supercoiled versus linear plasmid DNA and will give an indication about the DNA concentration.

3.2 *Rhodococcus fascians*

3.2.1 Large scale preparation of CCC plasmids in *R. fascians*

Covalently closed circular plasmids can be prepared from *R. fascians* on a large scale, essentially as described in Section 2.2.2, except that larger cultures are used, and the plasmid DNA is purified by CsCl equilibrium density gradient centrifugation, instead of by phenol extraction. *Protocol 9* yields 20 to 100 μg of plasmid DNA per litre of starting culture, depending mainly on the size of the plasmid.

Protocol 9. Large scale preparation of CCC plasmids from *R. fascians* (26)

- TRE buffer (see *Protocol 3*)
- solutions B and C (see *Protocol 2*)
- isopropanol
- CsCl
- ethidium bromide (10 mg/ml)
- TE saturated isopropanol
- ethanol (70%)

Method
1. Grow *R. fascians* cells in 2 litre YEB at 28°C (two days).
2. Harvest cells by centrifugation in fixed-angle rotor (e.g. Sorvall GS3) at 6000 *g*.
3. Prepare detergent-sensitive cells as described in *Protocol 3*.
4. Resuspend cells in 10 ml of TRE buffer.
5. Add 20 ml solution B. Allow to stand at room temperature for 10 min.
6. Add 15 ml of solution C. Mix gently by inversion and incubate on ice for 1 h.
7. Centrifuge at 12000 *g* in a fixed-angle rotor (e.g. SS34) in a Sorvall-type centrifuge.
8. Collect cleared supernatant in a plastic centrifuge tube (30 ml) and precipitate nucleic acids by addition of 0.7 volume isopropanol.
9. Pellet nucleic acids by centrifugation (7000 *g*).

Protocol 9. *Continued*

10. Briefly dry pellet in a vacuum desiccator and resuspend in 10 ml of TRE buffer.

11. Dissolve gently 10 g of CsCl in the DNA solution. Add 0.5 ml of a 10 mg/ml ethidium bromide solution.

12. Transfer to Beckman Quick-Seal™ tubes and centrifuge overnight at 48 000 r.p.m. in a Beckman NVT65 rotor, or for 48 h in a Beckman 50Ti rotor.

13. Illuminate the gradient with UV light and collect the lower (plasmid) band.

14. Extract the ethidium bromide gently by repeated extractions with isopropanol, saturated with TE.

15. Add two volumes of water and an equal volume of isopropanol, mix, and precipitate plasmid DNA in a swinging-bucket rotor, (e.g. Sorvall HB4) by centrifugation at 9000 *g*.

16. Wash the pellet with 70% ethanol and briefly dry in a vacuum desiccator. Resuspend the pellet in 100–200 µl of water.

3.2.2 Preparation of linear plasmids by elution from preparative PFGE gels

Large linear plasmids can not be enriched or separated from chromosomal DNA in the same way as CCC plasmids. However, separation of linear plasmids by CHEF electrophoresis (see Section 2.2.1), followed by elution from these gels, circumvents this problem. Several elution methods can be used: elution from CHEF gels made of low temperature-melting agarose, elution using chaotropic agents and silica matrices, and pulsed-field electro-elution. Elution with glass matrix (using, for instance, the Gene Clean II kit, Bio101 Inc., La Jolla, CA) gives higher yields, although the DNA is sheared into 15–20 kb fragments, whereas pulsed-field electroelution (see *Protocol 10*) results in lower yields of almost intact DNA (J. Desomer, unpublished data).

Protocol 10. Pulsed-field electroelution of linear plasmids (adapted from 27)

- Model 422 Electro Eluter™ (Bio-Rad, Cat. No. 165–2976)
- Pulse Wave™ 760 electrophoretic Field Switcher and Programming Block (FIGE) (Bio-Rad)
- isopropanol
- 10 × TAE buffer: 48.5 g Tris base, 11.4 ml acetic acid, 20 ml 0.5 M EDTA, pH 8.0, add H_2O to 1000 ml

Method

1. Separate linear plasmids via CHEF electrophoresis (see *Protocols 4* and *5*).

2. Push out agarose plugs containing the linear plasmids, using the glass tubes of the Electro Eluter™ system.

3. Cover the anode side of the glass tube with a dialysis membrane (M_r cut-off 12 000–15 000) and mount the tubes in the apparatus. Make sure that the tubes are completely filled with 1 × TAE buffer and remove all air bubbles trapped between agarose plugs and dialysis membrane. Several plugs can be placed in one glass tube.

4. Fill glass tubes and chamber with 1 × TAE buffer until the electrodes are covered. Place chamber in refrigerator.

5. Using a Pulse Wave Switcher™ (Bio-Rad) coupled to a power supply, electroelution is performed with the following FIGE parameters:

   • voltage: 175 V

   • forward versus inverse ratio 3:1

   • pulse time linear gradient from 6 sec to 147 sec

   • time: 6 h

6. Remove the liquid from the glass tubes and pool the solution that is trapped between plugs and dialysis membrane at the anode side.

7. Precipitate nucleic acids with 0.7 volume isopropanol.

8. Pellet the DNA and resuspend overnight in water or TE buffer.

4. Introduction of recombinant plasmids

4.1 *Agrobacterium tumefaciens*

4.1.1 *In vivo* techniques: mobilization

In most currently used procedures for introducing T-DNA vectors, *E. coli* is used as an intermediate cloning host for interspecies transfer. Non-self-transmissible plasmids, such as most T-DNA vectors, can nevertheless be transmitted if the required functions (*tra*, *mob*) are provided in *trans* from a separate (unrelated) plasmid. One of the best known examples is mobilization of ColE1 by F- or I-type plasmids. This process occurs at very high frequencies. At least one specific protein, encoded by the *mobility region* of ColE1, is essential for high frequency mobilization. This function can act in *trans* to complement pBR322-derived vectors which lack the *mobility region*. In addition to this *mob* function ColE1 carries a sequence called *bom* (basis of mobilization). Lack of this sequence results in non-mobilizability that can not be complemented in *trans*. pBR322-derived vectors still carry this *bom* site and can therefore be transferred by conjugation when the mob proteins are supplemented in *trans*.

Transfer of T-DNA vectors to *Agrobacterium* by mobilization is usually achieved by means of a triparental mating process, using the helper plasmid pRK2013. This plasmid contains the complete set of broad host range transfer functions from the IncP-type plasmid RK2, together with a kanamycin resistance marker cloned into ColE1 that also provides the *bom* functions (28). A typical triparental cross is described in *Protocol 11*.

Protocol 11. Triparental mating for mobilization of T-DNA vectors from *E. coli* to *Agrobacterium* (29)

1. Start three bacterial cultures in 3 ml of liquid LB medium, supplemented with the appropriate antibiotic: *E. coli* strain carrying the T-DNA vector, the helper *E. coli* strain HB101 (pRK2013), and the receptor *Agrobacterium* strain (e.g. C58C1Rif^r(pGV2260)).

2. Plate 0.1 ml of a 1:1:1 mixture of these strains on a LB agar plate and incubate overnight at 28°C.

3. Collect the mobilization mixture in 2 ml 10 mM MgSO$_4$ and plate serial dilutions (in 10 mM MgSO$_4$) on LB agar plates supplemented with antibiotics to select for the *Agrobacterium* strain that has received the T-DNA vector. Most T-DNA vectors contain a streptomycin resistance gene because it is an excellent marker in *Agrobacterium*.

4. Incubate the selection plates for two days at 28°C. Antibiotic resistant transconjugants are obtained at an average frequency of 10^{-5} per *Agrobacterium* recipient in the case of cointegration-type T-DNA vectors, or at 10^{-2} per recipient in the case of the binary vectors.

5. Streak transconjugants on selective LB agar plates to obtain single colonies.

6. Verify the presence of the mobilized plasmid in the colony purified transconjugant(s) as described in *Protocol 2*. When using cointegration-type vectors, prepare total bacterial DNA as described in *Protocol 17* and verify the cointegrate structure using a classical Southern blot hybridization technique.

4.2 *In vitro* techniques: high voltage electrotransformation

Until recently, no satisfactory way to introduce naked DNA directly into *Agrobacterium* was available. Using a freeze–thaw method, Holsters *et al.* (30) were able to obtain 10^3 transformed cells per microgram of DNA. Such low frequencies, as well as the alternative procedure of using *E. coli* as intermediate cloning host, were serious drawbacks for using *Agrobacterium* directly as a host for the construction of large libraries of genes. However, in recent years, high voltage-mediated electroporation has proved to be very useful for introducing naked DNA into cells, in particular into bacterial cells. A procedure is presented here for introducing plasmids directly into *Agrobacterium* by high voltage electroporation using the Gene-Pulser™ apparatus (Bio-Rad, Richmond CA, USA) equipped with a Pulse Controller™ and cuvettes with a 2 mm electrode gap distance. Efficiencies of 10^8 transformed cells per microgram of DNA are routinely obtained.

Protocol 12. Electrotransformation of *A. tumefaciens*

- microdialysis membranes (e.g. Millipore type VS, pore size 0.025 µm)

Method

1. Inoculate 250 ml liquid YEB medium with a saturated 5 ml pre-culture (grown overnight at 28°C and shaking at 250 r.p.m.).

2. Incubate the 250 ml of culture on a gyratory shaker at 250 r.p.m., until an OD_{600} of 0.3 is reached (∼ 6 h).

3. Collect the cells by centrifugation for 15 min at 6000 *g* and 4°C.

4. Wash the cells first in 250 ml and subsequently in 25 ml cold distilled water. Harvest the cells as described in step **3**.

5. Resuspend the bacteria in 500 µl cold distilled water.

6. Prior to electroporation, dialyse plasmid DNA against distilled water for at least 1 h. DNA can be stored at a concentration of 0.2 µg/µl.

7. Mix dialysed plasmid DNA with 400 µl cell suspension and leave on ice for 10 min.

8. Transfer the mixture to a pre-cooled electroporation cuvette.

9. Set the apparatus to a voltage of 12 500 V/cm, a resistance of 200 Ω, and a capacitance of 25 µF.

10. After the pulse, add 1 ml YEB medium to the cells, and incubate the bacteria for 2 h at 28°C with shaking at 250 r.p.m.

11. Plate appropriate dilutions on selective YEB agar plates and incubate for two days at 28°C.

12. Analyse transformants as described in Section 2.1.4.

Remark

Alternatively, washed cells can be resuspended in 15% PEG 6000 and frozen at −70°C for longer periods without drastically affecting transformation frequencies.

4.3 Introduction of recombinant plasmids in *Rhodococcus fascians* by electroporation

Although mobilization of recombinant plasmids between *E. coli* and *Corynebacterium* spp or *Mycobacterium* spp (31, 32) has been successfully achieved at high frequencies, no such phenomenon was observed when using *R. fascians* as acceptor strain (32). Equally, PEG-mediated uptake by protoplasts, as reported for other rhodococci (33–35), could not be achieved for *R. fascians* in our hands. Therefore, high voltage electrotransformation remains the preferred method for introducing recombinant plasmids into *R. fascians* (26). The most commonly used acceptor strain is a plasmid-less, streptomycin-resistant mutant strain D188-5 (obtained by growing a wild-type isolate at a sublethal temperature; 20). The protocol can be applied to all strains of *R. fascians* as well as to other rhodococci (e.g. *R. erythropolis*; J. Desomer, unpublished data). Preparation of electrocompetent cells of *R. fascians* (*Protocol 13*) involves only thorough washing and concentration of the cells in a medium with low conductivity. High voltage electroporations (see *Protocol 14*) were performed using a Gene-Pulser™ apparatus connected to a Pulse Controller™ (Bio-Rad Laboratories, Richmond, California) but other electroporation instruments gave similar transformation efficiencies (e.g. Cellject, Eurogentec, Belgium). No significant killing of the acceptor bacteria was observed using these conditions. Transformation efficiencies were dependent on the DNA concentration (ranging from 10^7/μg DNA at 1 ng/ml of DNA, to 10^5/μg DNA at 1 μg/ml of DNA) (26). These efficiencies allowed the isolation of recombinant plasmids by direct transformation of ligation mixtures to *R. fascians* as well as the rescue of non-replicating plasmids that integrate randomly in the genome of *R. fascians* (36; see also Section 5.2).

Different resistance genes are expressed in *R. fascians* and can be incorporated in cloning vectors (see *Table 3*). All the cloning vectors that have been described are based on the origin of replication of an endogenous *R. fascians* plasmid, pRF2 (located on a 3.5 kb *Xba*I–*Stu*I fragment; 36), which functions in other rhodococci as well (J. Desomer, K. Young, unpublished data). It remains to be established whether other rhodococcal plasmids that have been described (34, 37) can replicate in *R. fascians* and belong to the same incompatibility group as pRF2.

Protocol 13. Preparation of electrocompetent cells of *R. fascians*

- PEG 1000 (30%)

Method

1. Inoculate *R. fascians* in 50 ml of YEB medium, and allow to grow on a gyratory shaker at 28°C to OD$_{600}$ of 0.7–0.8.

2. Dilute the pre-culture in 0.5 litre of LB in a 2 litre Erlenmeyer and grow overnight at 28°C on a gyratory shaker.

3. Harvest the cells by centrifugation (9000 g).

4. Wash twice with ice-cold demineralized water.

5. Resuspend the cells in 10 ml 30% PEG 1000 solution in demineralized water (cell density $\sim 10^9$ c.f.u./ml).

6. Divide the cell suspension in 400 μl aliquots; store at −80°C.

Protocol 14. High voltage electrotransformation of *R. fascians* by plasmid DNA

1. Quickly thaw the frozen electrocompetent cells (see *Protocol 13*) and keep them on ice.

2. Mix 400 μl of ice-cold cells with the dialysed DNA of interest in a 2 mm gapped electrocuvette.

3. Deliver a 2.5 kV electric pulse from a 25 μF capacitor with the external resistance set at 400 Ω.

4. Dilute the pulsed cells immediately with 0.6 ml of YEB.

5. Incubate without shaking for 4 h at 28°C.

6. Spread cells on YEB plates containing appropriate antibiotics and incubate plates at 28°C.

7. Score transformants after five to seven days.

---

5. Analysis of plasmid-encoded functions

5.1 *Agrobacterium tumefaciens* Ti plasmids

Mutational analysis is the classical approach to elucidating the genetic and transcriptional organization of plasmid-encoded functions. In our laboratory, two main strategies for the mutagenesis of the Ti plasmid were followed.

The restriction site-directed approach relies on the availability of a detailed physical map; a specific plasmid fragment is cloned and restriction enzyme-directed insertions, substitutions, or deletions are generated. For insertions and substitutions, a restriction fragment encoding an antibiotic marker can be used, having the advantage that the mutation is tagged to a selective phenotype. However, this approach is limited by the availability of useful restriction sites. The problem can be overcome by performing a transposon saturation mutagenesis on a cloned Ti plasmid fragment, again tagging the mutation to a selective phenotype encoded by the transposon. In both cases, the mutated fragments are introduced into the appropriate *Agrobacterium* strain as

Table 3. Resistance markers and selective antibiotic and heavy metal concentrations for use with *R. fascians*

Antibiotic	Selective concentration μg/ml	Gene	Source	Reference
Chloramphenicol	25	*cmr*	*R. fascians*	41
Neomycin	5	*nptII*	Under control of *cmr* promoter	J. Desomer (unpublished data)
Kanamycin	25	*nptII*	Tn5	L. M. Mateos (unpublished data)
Thiostrepton	30	*tsr*	*Streptomyces*	
Phleomycin	1	*bat*	*Streptoverticillus*	26
Hygromycin	30	*hyg*	*Streptomyces hygroscopicus*	
Streptomycin	100	—	*R. fascians* spontaneous mutant	20
Cadmium nitrate	27.5	—	*R. fascians* plasmids	20

described in Section 4.1. Mutant Ti plasmids are produced by the exchange of the modified Ti plasmid segment for the wild-type sequence through a double homologous recombination on either side of the transposon. Two types of replicons can be used for this approach. In the first case, cloning and mutagenesis are carried out in a broad host range cloning vector that can replicate in *Agrobacterium* (e.g. a pVS1-derived plasmid). Recombinants between the non-conjugative intermediate vector and the resident Ti plasmid can be isolated after a Ti plasmid conjugation as described by Petit *et al.* (38).

The second and most useful method relies on suicide vectors for *Agrobacterium*, such as pBR-derived plasmids. Upon introduction into *Agrobacterium*, intermediate vectors of this type can only be maintained after cointegration into the resident Ti plasmid through homologous recombination between the corresponding regions of the Ti plasmid. The desired mutant Ti plasmid is obtained after a double recombination event, resulting in the loss of the cloning vector and thus the selective marker it encodes. This second method has the great advantage that non-conjugative plasmids or even entire genomes can also be mutagenized.

In cases where there are no physical data available for a particular (Ti) plasmid, a random transposon mutagenesis experiment can be performed. For this purpose a transposon (e.g. Tn5) carried by a suicide vector (e.g. pBR322 or pRK2013) can be introduced into *Agrobacterium* via mobilization (see Section 4.1.1) or electroporation (see Section 4.1.2) and selection for the transposon-encoded marker. A subsequent Ti plasmid conjugation experiment as described above, allows discrimination between transpositions to the Ti plasmid or to the chromosome.

5.2 *Rhodococcus fascians* plasmids

To analyse plasmid functions in *R. fascians* the insertion mutagenesis method based on illegitimate recombination (36) can be used. Therefore, pUC-derived vectors containing a marker gene (e.g. pRF41 containing the gene coding for chloramphenicol resistance *cmr*) are introduced into the *R. fascians* strain containing the plasmid of interest by high voltage electrotransformation (see Section 4.2). These plasmids are unable to replicate, but Cm^r transformants can be rescued by integration in the genome of *R. fascians*. This illegitimate integration occurs almost randomly along the genome, generating auxotrophic or pigmentation mutants with high frequency. The recombination sites in the plasmid, however, are limited to sequences overlapping the *Nar*I restriction site, resulting in integrated plasmids with similar structure in all mutants (see *Figure 2*). The presence of a ColE1 replicon and the Ap^r marker gene allows straightforward isolation in *E. coli* of the inserted sequences and adjacent *R. fascians* DNA from a mutant strain of interest (*Protocol 16*).

pRF41 insertions in conjugative *R. fascians* plasmids can be enriched by

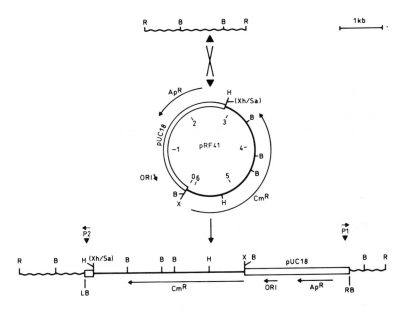

Figure 2. Schematic representation of pRF41 as a plasmid and upon integration. The *wavy lines* indicate *R. fascians* chromosomal sequences and the *solid lines* represent pRF28 sequences. The pUC18 sequence is shown as an *open box*. Ori, origin of replication of pUC18; Ap^R, ampicillin resistance gene; Cm^R, chloramphenicol resistance gene; RB, right boundary; LB, left boundary; P1 and P2, oligodeoxynucleotides used as primers for dideoxy chain termination sequencing (36); R, restriction enzymes that do not cleave pRF41. B, *Bam*HI; E, *Eco*RI; X, *Xba*I; H, *Hin*dIII; Xh, *Xho*I; Sa, *Sal*I. Brackets indicate destruction of a restriction site by ligation of a fragment which has compatible sticky ends.

conjugation between pooled Cm^r transformants and the cured Sm^r acceptor strain D188-5 (see *Protocol 15*). In this way, several loci on pFiD188 involved in fasciation have been identified (1).

Protocol 15. Conjugation between *R. fascians* strains (20)

- Millipore nitrocellulose filter (0.2 μm pore size)
- $MgSO_4$ (10 mM)

Method

1. Grow *R. fascians* cultures at 28°C in liquid YEB medium until early stationary phase.
2. Mix acceptor and donor strains in a 1:1 ratio (usually 5 ml of each culture).
3. Filter 5 ml of the bacterial suspension and recover the filter.
4. Incubate the filter, with the bacteria on top, on a YEB agar plate for two days at 28°C.
5. Resuspend in 5 ml buffer containing 10 mM $MgSO_4$.
6. Plate appropriate dilutions (10^0, 10^{-2}, 10^{-4}, and 10^{-6} on YEB plates selective for acceptor bacteria, donor bacteria, and for transconjugants (containing both selective agents), respectively. Controls with only acceptor or donor bacteria must be included to estimate mutation frequencies.

Remark

Frequencies range from 10^{-7} to 10^{-2} for CCC plasmids and are about 10^{-4} for Fi plasmids.

Protocol 16. Construction of *E. coli* clones containing *R. fascians* DNA adjacent to pRF41 insertions from mutant strains (36)

1. Prepare total DNA from the *R. fascians* mutant strain (see *Protocol 18*).
2. Digest 1 to 2 μg of total DNA to completion with *Bam*HI or *Bgl*II.
3. Phenol extract the restricted DNA, precipitate, and resuspend in 50 μl ligation buffer.
4. Add one unit T4 DNA ligase and incubate at room temperature for 4 h.
5. Heat inactivate the ligase for 10 min at 65°C and use half of the mixture to transform competent *E. coli* MC1061.

6. Select on LB agar plates supplemented with a β-lactam antibiotic, e.g. tricarcillin 100 μg/ml (carbenicillin or ampicillin can also be used).

Remark
*Bg*lll does not recognize any sequences within pRF41 and therefore yields clones that contain both left and right junction fragments. *Bam*HI generated clones normally contain only the right border sequences.

5.3 Analysis of insertion mutants via Southern-type hybridizations

The physical characterization of mutations, insertions, or cointegrate structures on the Ti or Fi plasmids by restriction enzyme analysis of isolated plasmid DNA is quite time-consuming. The following protocols provide a simple and rapid alternative for characterizing mutant Ti or Fi plasmids. It involves the isolation of total bacterial DNA followed by a Southern blotting hybridization of digested DNA.

Protocol 17. Isolation of total *Agrobacterium* DNA (39)

- TE buffer (see *Protocol 2*)
- Sarkosyl
- Pronase
- Tris–HCl-saturated phenol
- chloroform
- NaCl
- ethanol

Method

1. Grow *Agrobacterium* cultures overnight in 3 ml LB medium at 28 °C.
2. Transfer 1.5 ml of the culture in an Eppendorf tube and collect the cells by centrifugation for 2 min at 13 000 r.p.m.
3. Resuspend the pellet in 300 μl TE buffer.
4. Add 100 μl 5% Sarkosyl (in TE buffer).
5. Add 100 μl Pronase solution (2.5 mg/ml in TE buffer).
6. Incubate at 37 °C for 1 h or more until complete lysis is achieved.
7. Shear the lysate by several passages through a syringe (with 20 gauge needle); the lysate will become less viscous.
8. Extract twice with an equal volume of Tris–HCl saturated phenol.
9. Extract residual phenol twice with chloroform.
10. Precipitate the nucleic acid by adding NaCl to 0.25 M and two volumes ethanol, and put at −20 °C for 1 h.
11. Centrifuge the pellet for 5 min.

Protocol 17. *Continued*

12. Discard the supernatant, dry the pellet, and dissolve in 100 μl H_2O.

13. Measure the UV spectrum to determine the DNA concentration, and digest 5 μg of this total DNA.

14. Upon agarose gel electrophoresis, proceed with a classical Southern blotting hybridization experiment (40).

Protocol 18. Total DNA preparation from *R. fascians* (36)

- TRE buffer (see *Protocol 3*)
- PEG 6000
- lysozyme
- SDS
- Tris–HCl-equilibrated phenol
- Tris–HCl-equilibrated diethyl ether
- ethanol

Method

1. Grow *R. fascians* cultures in 50 ml at 28°C on an orbital shaker for two days.

2. Harvest cells by centrifugation (e.g. Sorvall SS34 rotor, 7000 *g*).

3. Resuspend cells in 5 ml PEG 6000 (20%) in TRE buffer, containing 1 mg/ml lysozyme.

4. Incubate 2 h at 37°C.

5. Harvest cells and resuspend in 1 ml TRE buffer.

6. Add 1 ml 1% SDS to accomplish lysis. Mix gently and allow to stand for 10 min at room temperature.

7. Extract the lysate twice with Tris–HCl (pH 7.2)-equilibrated phenol.

8. Extract residual phenol with Tris–HCl-equilibrated diethyl ether.

9. Precipitate nucleic acids with ethanol.

10. Precipitated DNA can usually be isolated with a glass rod.

11. Briefly dry DNA and resuspend in H_2O.

Acknowledgements

The authors would like to thank M. De Cock, V. Vermaercke, and K. Spruyt for the assistance in preparing the manuscript and pictures. J.D. is a post-doctoral researcher of the National Fund for Scientific Research (Belgium). Research was supported by the Services of the Prime Minister (IUAP 12 OCU 192) and 'Vlaams Actieprogramma Biotechnologie' (174 KP 490).

Table 4. Bacterial media

Solid media contain 15 g agar per litre

LB medium

Bactotryptone	10 g/litre
Bacto yeast extract	5 g/litre
NaCl	10 g/litre
pH 7.2	

YEB medium

Bacto beef extract	5 g/litre
Bacto yeast extract	1 g/litre
Peptone	5 g/litre
Sucrose	5 g/litre
$MgSO_4$	2 mM
pH 7.2	

PA medium

Peptone	4 g/litre
$MgSO_4$	2 mM
pH 7.2	

Minimal medium

K_2HPO_4	10.5 g/litre
KH_2PO_4	4.5 g/litre
$(NH_4)_2SO_4$	1.0 g/litre
Sodium citrate.$2H_2O$	0.5 g/litre

These salts are added as a 5 × stock to the agar solution.
Add to the autoclaved medium:
1ml 20% $MgSO_4.7H_2O$
10 ml 20% glucose.

References

1. Crespi, M., Messens, E., Caplan, A. B., Van Montagu, M., and Desomer, J. (1992). *EMBO J.*, **11**, 795.
2. Van den Eede, G., Deblaere, R., Goethals, K., Van Montagu, M., and Holsters, M. (1992). *Mol. Plant-Microbe Interactions*, **5**, 228.
3. Klee, H. J. and Rogers, S. G. (1989). In *Molecular biology of plant nuclear genes* (Cell culture and somatic cell genetics of plants) (ed. J. Schell and I.K. Vasil), Vol. 6, pp. 2–25. Academic Press, San Diego.
4. Depicker, A. and Gheysen, G. (1991). In *Biochemical aspects of crop improvement* (ed. K. R. Khanna), pp. 399–420. CRC Press, Boca Raton.
5. Zambryski, P. (1992). *Annu. Rev. Plant Physiol. Plant Mol. Biol.*, **43**, 465.
6. Tempé, J. and Petit, A. (1983). In *Molecular genetics of the bacteria-plant interaction* (ed. A. Pühler), pp. 14–32. Springer-Verlag, Berlin.
7. Akiyoshi, D. E., Klee, H., Amasino, R. M., Nester, E. W., and Gordon, M. P. (1984). *Proc. Natl. Acad. Sci. USA*, **81**, 5994.
8. Schröder, G., Waffenschmidt, S., Weiler, E. W., and Schröder, J. (1984). *Eur. J. Biochem.*, **138**, 387.
9. Zambryski, P., Joos, H., Genetello, C., Leemans, J., Van Montagu, M., and Schell, J. (1983). *EMBO J.*, **2**, 2143.
10. Deblaere, R., Reynaerts, A., Höfte, H., Hernalsteens, J.-P., Leemans, J., and Van Montagu, M. (1987). In *Methods in enzymology* (ed. R. Wu and L. Grossman), Vol. 153, pp. 277–92. Academic Press, New York.
11. Hoekema, A., Hirsch, P. R., Hooykaas, P. J. J., and Schilperoort, R. A. (1983). *Nature*, **303**, 179.
12. Deblaere, R., Bytebier, B., De Greve, H., Deboeck, F., Schell, J., Van Montagu, M., and Leemans, J. (1985). *Nucleic Acids Res.*, **13**, 4777.

13. LeChevalier, H. A. (1986). In *Bergey's manual of systematic bacteriology* (ed. P. H. A. Sneath, N. S. Mair, M. E. Sharpe, and J. G. Holt), Vol. 2, pp. 1458–506. Williams & Wilkins, Baltimore.
14. Brown, N. A. (1927). *Phytopathology*, **17**, 29.
15. Tilford, P. E. (1936). *J. Agric. Res.*, **53**, 383.
16. Stapp, C. (ed.) (1961). *Bacterial plant pathogens*. Oxford University Press, Oxford, UK.
17. Finnerty, W. R. (1992). *Annu. Rev. Microbiol.*, **46**, 193.
18. Murai, N., Skoog, F., Doyle, M. E., and Hanson, R. S. (1980). *Proc. Natl. Acad. Sci. USA*, **77**, 619.
19. Lawson, E. N., Gantotti, B. V., and Starr, M. P. (1982). *Curr. Microbiol.*, **7**, 327.
20. Desomer, J., Dhaese, P., and Van Montagu, M. (1988). *J. Bacteriol.*, **170**, 2401.
21. Sakaguchi, K. (1990). *Microbiol. Rev.*, **54**, 66.
22. Eckhardt, T. (1978). *Plasmid*, **1**, 584.
23. Chassy, B. M. (1976). *Biochem. Biophys. Res. Commun.*, **68**, 603.
24. Birnboim, H. C. and Doly, J. (1979). *Nucleic Acids Res.*, **7**, 1513.
25. Hirsch, P. R., Van Montagu, M., Johnston, A. W. B., Brewin, N. J., and Schell, J. (1980). *J. Gen. Microbiol.*, **120**, 403.
26. Desomer, J., Dhaese, P., and Van Montagu, M. (1990). *Appl. Environm. Microbiol.*, **56**, 2818.
27. Goto, K., Motoyoshi, T., Tamura, G., Obata, T., and Hara, S. (1990). *Agric. Biol. Chem.*, **54**, 1499.
28. Figurski, D. H. and Helinski, D. R. (1979). *Proc. Natl. Acad. Sci. USA*, **76**, 1648.
29. Ditta, G., Stanfield, S., Corbin, D., and Helinski, D. R. (1980). *Proc. Natl. Acad. Sci. USA*, **77**, 7347.
30. Holsters, M., De Waele, D., Depicker, A., Messens, E., Van Montagu, M., and Schell, J. (1978). *Mol. Gen. Genet.*, **163**, 181.
31. Gormley, E. P. and Davies, J. (1991). *J. Bacteriol.*, **173**, 6705.
32. Schäfer, A., Kalinowski, J., Simon, R., Seep-Feldhaus, A.-H., and Pühler, A. (1990). *J. Bacteriol.*, **172**, 1663.
33. Brownell, G. H., Saba, J. A., Denniston, K., and Enquist, L. W. (1982). *Dev. Ind. Microbiol.*, **23**, 287.
34. Vogt Singer, M. E. and Finnerty, W. R. (1988). *J. Bacteriol.*, **170**, 638.
35. Dabbs, E. R. and Sole, G. J. (1988). *Mol. Gen. Genet.*, **211**, 148.
36. Desomer, J., Crespi, M., and Van Montagu, M. (1991). *Mol. Microbiol.*, **5**, 2115.
37. Hashimoto, Y., Nishiyama, M., Yu, F., Watanabe, I., Horinouchi, S., and Beppu, T. (1992). *J. Gen. Microbiol.*, **138**, 1003.
38. Petit, A., Dessaux, Y., and Tempé, J. (1978). In *Proceedings IVth International Conference on Plant Pathogenic Bacteria* (ed. M. Ridé), pp. 143–52. I.N.R.A., Angers.
39. Dhaese, P., De Greve, H., Decraemer, H., Schell, J., and Van Montagu, M. (1979). *Nucleic Acids Res.*, **7**, 1837.
40. Sambrook, J., Fritsch, E. F., and Maniatis, T. (ed.) (1989). *Molecular cloning, a laboratory manual*. Cold Spring Harbor Laboratory Press, NY.
41. Desomer, J., Vereecke, D., Crespi, M., and Van Montagu, M. (1992). *Mol. Microbiol.*, **6**, 2377.

6

Streptomyces plasmid vectors

W. WOHLLEBEN and G. MUTH

1. Introduction

Bacteria of the genus *Streptomyces* are Gram-positive, mycelium-forming soil bacteria that are generally non-pathogenic and which comprise a large number of different species. They are classified according to the following criteria: morphological (e.g. colour and shape of the mycelium and spores), physiological (e.g. utilization of different nitrogen and carbon sources, production of primary and secondary metabolites), and genetic (e.g. rRNA sequences) (1, 2).

A common feature of all *Streptomyces* species is a morphologically complex life cycle. During vegetative growth, substrate mycelium develops out of germinating spores. Thereafter, generative growth starts with aerial mycelium formation which is completed by the generation of new spores. Cellular differentiation and its multilevel regulation are interesting fields of research studied by several groups.

Morphological differentiation is accompanied by physiological differentiation, resulting in the production of secondary metabolites. The large number of *Streptomyces* strains synthesize a multitude of secondary metabolites which are of medical and technical interest. Streptomycetes have been used for industrial fermentation for more than 40 years, especially for antibiotic production.

In addition, streptomycetes exhibit exceptional nutritional versatility based on their ability to hydrolyse complex organic compounds in their environment. A variety of extracellular enzymes (such as amylases, cellulases, proteases, and nucleases) has been characterized (3).

For the analysis of *Streptomyces* on the molecular level various methods have been developed, many of which have been summarized in a laboratory manual (4). The application of molecular techniques has apparently not yet led to dramatic results in industrial strain improvement programmes. But these methods have provided a lot of information, in particular on differentiation and antibiotic biosynthesis. A great variety of genes involved in antibiotic biosynthesis has been cloned (for a review see 5) and much has been learnt about the molecular analysis of primary metabolism, exoenzyme production, and in particular on regulatory phenomena involved in switching the bacteria from primary to secondary metabolism.

All this knowledge may lead to the over-production of *Streptomyces* gene products on an industrial scale in the near future, and may be helpful for the generation of new, antibiotically active compounds (such as hybrid antibiotics (6)) by genetic engineering (7).

Streptomyces strains can also be used for the secretion of heterologous proteins. Following fusion to signal sequences, various proteins have been secreted by streptomycetes (8).

The standard molecular techniques for *Streptomyces* have been developed mainly for *S. coelicolor* and *S. lividans*, the commonly used *Streptomyces* strains. Most of these methods can be adapted to other *Streptomyces* strains after some modifications (as indicated in the protocols).

The development of cloning techniques began with the discovery of plasmids in streptomycetes and the establishment of a transformation procedure (9) using polyethylene glycol (PEG). Since many *Streptomyces* strains harbour plasmids (see Section 2), a variety of basic replicons are available, some of which have proven to be suitable for vector construction.

In the past few years, emphasis in this field has shifted from the development of 'classical' vectors for cloning and expression to the construction of vectors for reverse cloning and reverse genetics. In this chapter we summarize the different vector systems and describe the application of non-replicative and temperature-sensitive vectors for reverse cloning and reverse genetics.

2. Autonomously replicating Streptomyces vectors

2.1 Properties of *Streptomyces* plasmids

Four distinct classes of plasmids have been isolated from *Streptomyces* strains.

(a) Large low copy number plasmids which are very stably maintained in host cells.

(b) Small multicopy plasmids which represent the majority of plasmids found in streptomycetes and replicate via the rolling circle mechanism.

(c) Extrachromosomal elements arising by excision from the chromosome that can replicate autonomously in certain strains but which can also integrate into the host chromosome via site-specific recombination.

(d) Giant linear plasmids which often encode complete antibiotic production pathways (10).

A property common to most of these different plasmids is their self-transmissibility. Plasmids are transferred by conjugation at frequencies of up to 100%. Plasmid transfer is often accompanied by the mobilization of chromosomal markers. It is normally associated with the formation of pocks, characteristic inhibition zones where the growth of the aerial mycelium of the recipient is retarded (11). In contrast to the transfer systems of plasmids from Gram-negative bacteria, only very few plasmid-encoded transfer functions are required.

With the exception of the linear plasmids that are several 100 kb in size, plasmids from each class have been used for the development of cloning vectors (12) and in particular the following.

2.1.1 SCP2*

Plasmid SCP2 (and its high fertility mutant SCP2*, respectively) was the first *Streptomyces* plasmids to be isolated (13, 14). By deletion and insertion analysis the DNA regions essential for replication, copy number control, self-transmissibility, and stable maintenance have been characterized. A series of cloning vectors derived from SCP2* were found to replicate very stably and to maintain a copy number of about one per cell (15). The most important property of the SCP2-based cloning vectors is their ability to accept very large DNA inserts of up to more than 30 kb. Thus, by using SCP2-derived vectors even complete antibiotic biosynthetic pathways can be cloned and transferred into heterologous hosts (16).

2.1.2 SLP1

The SLP1 family (SLP1.1, SLP1.2, SLP1.3, etc.) represents a set of deletion derivatives of the SLP1 extrachromosomal element (17). Smaller deletion derivatives of plasmid SLP1, lacking the integrase gene (see below) were the basis of the early vectors pIJ61 or pIJ41 (18). These vectors have a copy number of about five and a narrow host range.

2.1.3 pIJ101

The most widely used *Streptomyces* vectors have been derived from the small multicopy *S. lividans* plasmid pIJ101 (19, 20). The organization of plasmid pIJ101, which is typical for a multicopy *Streptomyces* plasmid, is shown in *Figure 1*.

(a) The Rep protein initiates replication of the leading strand at the origin.

(b) The *sti* (strong incompatibility) function represents the minus origin for initiation of lagging strand synthesis. Plasmids that lack the *sti* function accumulate large amounts of single-stranded (ss) DNA and are not compatible with pIJ101 derivatives that carry the *sti* region.

(c) The *tra* (KilA) function, which is regulated at the transcriptional level by the KorA repressor, is responsible for intermycelial plasmid transfer.

(d) The spread (*spd*) function mediates the intramycelial transfer of the plasmid.

4.1.1 Gene disruption experiments using temperature-sensitive plasmids

If the disruption experiments are performed with pGM vectors, the choice of the corresponding pGM vector depends on both the restriction sites and the endogenous antibiotic resistance of the strain (*Table 2, Figure 4*). The experiment is then as follows:

(a) The vector is introduced into the respective strain (*Protocol 1*).

(b) The most efficient conditions are determined for eliminating the vector molecules lacking inserts (*Protocol 2*).

(c) The disruption experiment can then be performed using the plasmid which carries the insert (*Protocol 3*).

Protocol 1. PEG transformation of *Streptomyces* protoplasts [a]

Materials

- S medium: 4 g peptone, 4 g yeast extract, 4 g K_2HPO_4, 2 g KH_2PO_4, in 795 ml H_2O, 200 ml 5% glucose, 5 ml 10% $MgSO_4$, prepare and autoclave separately

- P buffer: 10.3% sucrose, 25 mM TES (adjusted with NaOH, pH 7.2), 10 mM $MgCl_2$, 2.5 mM $CaCl_2$, 1.4 mM K_2SO_4, 0.4 mM KH_2PO_4, 0.2% trace element solution [a]

- T buffer: 2.5% sucrose, 50 mM Tris–maleate, 10 mM $MgCl_2$, 10 mM $CaCl_2$, 0.4 mM KH_2PO_4, 1.4 mM K_2SO_4, 25% PEG 1000 (Serva, Heidelberg, Germany), 0.2% trace element solution, [b] pH 8.0

- R2YE plates: 103 g sucrose, 10 g glucose, 0.25 g K_2SO_4, 10.12 g $MgCl_2$, 0.1 g casamino acids (Difco), 5 g yeast extract, 5.73 g TES, 17 g agar in 900 ml H_2O, 80 ml 3.68% $CaCl_2$, 10 ml 0.5% KH_2PO_4, 15 ml 20% L-proline, 8 ml 1 N NaOH, 0.2 ml trace element solution, prepare and autoclave separately

- NB soft agar: 0.8% nutrient broth, 0.5% agar

A. *Protoplasting of streptomycetes*

1. Homogenize 35 ml of a two day *Streptomyces* culture that has grown well, cultivated in S medium supplemented with glycine. [c]

2. Spin down in a refrigerated high speed centrifuge (10 000 r.p.m., 10 min). [d]

3. Wash in P buffer.

4. Resuspend pellet in 5 ml P buffer containing lysozyme (2 mg/ml).

5. Triturate with a 20 ml pipette to mix thoroughly.

6. Incubate for 15–120 min at 30°C until most of the cells are protoplasts (check with a microscope). [e]

7. Add 10 ml P buffer.

8. Triturate to release protoplasts from mycelium.

9. Filter through cotton wool to separate non-protoplasted mycelium from protoplasts.

10. Wash protoplasts in P buffer.

11. Sediment protoplasts by spinning in a refrigerated high speed centrifuge[d] (5000 r.p.m., 10 min).

12. Discard supernatant and resuspend pellet in 1–5 ml P buffer (10^9–10^{10} protoplasts/ml).

13. Use protoplasts directly for transformation or freeze aliquots for storage at $-20\,^{\circ}$C.

B. *PEG transformation*

1. Mix 100 (200) μl protoplasts (10^8–10^9) with 5–20 μl plasmid DNA (0.1–1 μg/μl) and 250 (500) μl PEG-containing T buffer.

2. Immediately spread 100 μl on each of six to eight pre-dried R2YE plates (use a pipette for spreading).[f]

3. Incubate at 28 °C for 16–20 h.

4. Overlay with 2.5 ml NB soft agar containing appropriate antibiotic concentrations (*Table 1*).

5. Incubate at 28 °C. Resistant colonies, (i.e. transformants) can normally be detected after about two days.

[a] Modification of the method of Hopwood *et al.* (4).
[b] Autoclave separately, compounds see *Table 3*.
[c] Glycine concentration depends on strains to be transformed, e.g. 1.0% for *S. lividans*; 0.75% for *S. viridochromogenes*.
[d] Bench centrifuges may be used instead of a refrigerated high speed centrifuge.
[e] Incubation time depends on the strains, e.g. 30 min for *S. lividans*, 15 min for *S. viridochromogenes*.
[f] For some strains it is necessary to embed the protoplasts in soft agar for efficient regeneration.

Table 3. Trace element solution (in 1 litre H_2O)

200 mg $FeCl_3.6H_2O$
10 mg $Na_2B_4O_7.10H_2O$
10 mg $(NH_4)_6Mo_7O_{24}.4H_2O$
10 mg $CuCl_2.2H_2O$
10 mg $MnCl_2.4H_2O$
40 mg $ZnCl_2$

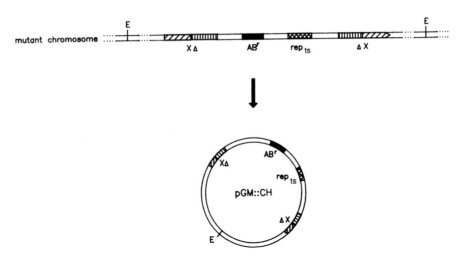

Figure 9. Mutational cloning scheme (60). Gene disruption mutants (*Figure 6*) can be used to identify DNA fragments adjacent to the mutated gene. Following isolation of genomic DNA from the mutant, the DNA is cleaved with a restriction enzyme (E) which does not cleave the plasmid originally used. The religation, transformation into a suitable recipient, and selection for the vector marker (AB^r) results in a hybrid plasmid (pGM:CH) which carries DNA fragments flanking gene X.

hybridize to them can then be used for the identification and inactivation of the desired gene (reverse genetics).

Acknowledgements

We thank M. Guérineau for communicating results prior to publication and all members of the 'Streptomyces-group' of our department who helped to work out and to test the different plasmid suicide systems. In particular, we are grateful to D. Hillemann, G. Labes, and S. Pelzer who developed some of the protocols included in this chapter. We also thank K. Krey for preparing illustrations and M. Labes for her critical reading of the manuscript. The authors' research work was supported by the Deutsche Forschungsgemeinschaft (Wo 485/1-1) and the Bundesministerium für Forschung und Technologie (0319374A).

References

1. Langham, C. D., Williams, S. T., Sneath, P. H. A., and Mortimer, A. M. (1989). *J. Gen. Microbiol.*, **135**, 121.
2. Kämpfer, P., Kroppenstedt, R. M., and Dott, W. (1991). *J. Gen. Microbiol.*, **137**, 1831.

3. Deshpande, B. S., Ambedkar, S. S., and Shewale, J. G. (1988). *Enzyme Microb. Technol.*, **10**, 455.
4. Hopwood, D. A., Bibb, M. J., Chater, K. F., Kieser, T., Bruton, C. J., Kieser, H. M., Lydiate, D. J., Smith, C. P., Ward, J. M., and Schrempf, H. (1985). *Genetic manipulation of streptomyces: a laboratory manual.* John Innes Foundation, Norwich.
5. Chater, K. F. (1990). *Bio/Technology*, **8**, 115.
6. Hopwood, D. A., Malpartida, F., Kieser, H. M., Ikeda, H., Duncan, J., Fujii, I., Rudd, B. A. M., Floss, H. G., and Omura, S. (1985). *Nature*, **314**, 642.
7. Hopwood, D. A. (1989). *Phil. Trans. Roy. Soc. Lond.*, **324**, 549.
8. Engels, J. W. and Koller, K.-P. (1991). In *Transgenesis, application of gene transfer* (ed. J. A. H. Murray), pp. 33–53. University Press, Buckingham.
9. Bibb, M. J., Schottel, J. L., and Cohen, S. N. (1980). *Nature*, **284**, 526.
10. Kinashi, H., Shimaji, H., and Sakai, A. (1987). *Nature*, **328**, 454.
11. Hopwood, D. A., Lydiate, D. J., Malpartida, F., and Wright, H. M. (1985). *Basic Life Sciences*, **30**, 615.
12. Hopwood, D. A., Kieser, T., Lydiate, D. J., and Bibb, M. J. (1986). In *The bacteria* (ed. S. W. Queener and L. E. Day), Vol. IX, pp. 159–230. Academic Press, Orlando.
13. Bibb, M. J., Freeman, R. F., and Hopwood, D. A. (1977). *Mol. Gen. Genet.*, **154**, 155.
14. Schrempf, H., Bujard, H., Hopwood, D. A., and Goebel, W. (1975). *J. Bacteriol.*, **121**, 416.
15. Lydiate, D. J., Malpartida, F., and Hopwood, D. A. (1985). *Gene*, **35**, 223.
16. Malpartida, F. and Hopwood, D. A. (1984). *Nature*, **309**, 462.
17. Bibb, M. J., Ward, M. J., Kieser, T., Cohen, S. N., and Hopwood, D. A. (1981). *Mol. Gen. Genet.*, **184**, 230.
18. Thompson, C. J., Kieser, T., Ward, J. M., and Hopwood, D. A. (1982). *Gene*, **20**, 51.
19. Kieser, T., Hopwood, D. A., Wright, H. M., and Thompson, C. J. (1982). *Mol. Gen. Genet.*, **185**, 223.
20. Kendall, K. J. and Cohen, S. N. (1988). *J. Bacteriol.*, **170**, 4634.
21. Gormley, E. P. and Davies, J. (1991). *J. Bacteriol.*, **173**, 6705.
22. Jorgensen, R. A., Rothstein, S. J., and Reznikoff, W. S. (1979). *Mol. Gen. Genet.*, **177**, 65.
23. Wohlleben, W., Arnold, W., Bissonnette, L., Pelletier, A., Tanguay, A., Roy, P. H., Gamboa, G. C., Barry, G. F., Aubert, E., Davies, J., and Kagan, S. A. (1989). *Mol. Gen. Genet.*, **217**, 202.
24. Ingram, C., Brawner, M., Youngman, P., and Westpheling, J. (1989). *J. Bacteriol.*, **171**, 6617.
25. Schauer, A., Ranes, M., Santamaria, R., Guijarro, J., Lawlor, E., Mendez, C., Chater, K. F., and Losick, R. (1989). *Science*, **240**, 768.
26. Thompson, C. J., Ward, J. M., and Hopwood, D. A. (1980). *Nature*, **286**, 525.
27. Malpartida, F., Zalacain, M., Jimenez, A., and Davies, J. (1983). *Biochem. Biophys. Res. Commun.*, **117**, 6.
28. Gil, J. A., Kieser, H. M., and Hopwood, D. A. (1985). *Gene*, **38**, 1.
29. Drocourt, D., Calmels, T., Reynes, J. P., Baron, M., and Tiraby, G. (1990). *Nucleic Acids Res.*, **18**, 4009.

30. Berg, D. E. (1989). In *Mobile DNA* (ed. D. E. Berg and M. M. Howe), pp. 185–210. ASM, Washington DC.

31. Stanzak, R., Matsushima, P., Baltz, R. H., and Rao, R. N. (1986). *Biotechnology*, **4**, 229.

32. Katz, E., Thompson, C. J., and Hopwood, D. A. (1983). *J. Gen. Microbiol.*, **129**, 2703.

33. Hopwood, D. A., Bibb, M. J., Chater, K. F., and Kieser, T. (1987). In *Methods in enzymology* (ed. R. Wu and L. Grossman), Vol. 153, pp. 116–65.

34. Bibb, M. J. and Cohen, S. N. (1982). *Mol. Gen. Genet.*, **187**, 265.

35. Jaurin, B. (1987). *Nucleic Acids Res.*, **15**, 8567.

36. Davis, N. K. and Chater, K. F. (1990). *Mol. Microbiol.*, **4**, 1679.

37. Horinouchi, S. and Beppu, T. (1985). *J. Bacteriol.*, **162**, 406.

38. Ward, J. M., Janssen, G. R., Kieser, T., Bibb, M. J., Buttner, M. J., and Bibb, M. J. (1986). *Mol. Gen. Genet.*, **203**, 468.

39. Hahn, D. R., Solenberg, P. J., and Baltz, R. H. (1991). *J. Bacteriol.*, **173**, 5573.

40. Solenberg, P. J. and Burgett, S. G. (1989). *J. Bacteriol.*, **171**, 4807.

41. Sohaskey, C. D., Im, H., and Schauer, A. T. (1992). *J. Bacteriol.*, **174**, 367.

42. Chung, S. T. (1987). *J. Bacteriol.*, **169**, 4436.

43. Bibb, M. J. and Janssen, G. R. (1986). In *Fifth International Symposium on the Genetics of Industrial Micro-organisms* (ed. M. Alacevic, D. Hranueli, and Z. Toman), pp. 309–18. Ognjen Prica Printing Works, Karlovac.

44. Murakami, T., Holt, T. G., and Thompson, C. J. (1989). *J. Bacteriol.*, **171**, 1459.

45. Kuhstoss, S. and Rao, N. (1991). *Gene*, **103**, 97.

46. Smokvina, T., Mazodier, P., Boccard, F., Thompson, C. J., and Guérineau, M. (1990). *Gene*, **94**, 53.

47. Kieser, T. and Melton, R. (1988). *Gene*, **65**, 83.

48. Rao, N. R., Richardson, M. A., and Kuhstoss, S. (1987). In *Methods in enzymology* (ed. R. Wu and L. Grossman), Vol. 153, pp. 167–99.

49. Portmore, J. J., Burnett, W. V., and Chiang, S.-J. (1987). *Biotechnol. Lett.*, **9**, 7.

50. Chater, K. F. (1986). In *The bacteria* (ed. S. W. Queener and L. E. Day), Vol. IX, pp. 119–58. Academic Press, Orlando.

51. Pernodet, J. L., Simonet, J. M., and Guérineau, M. (1984). *Mol. Gen. Genet.*, **198**, 35.

52. Mazodier, P., Thompson, C., and Boccard, F. (1990). *Mol. Gen. Genet.*, **222**, 431.

53. Martin, C., Mazodier, P., Mediola, M. V., Giquel, B., Smokvina, T., Thompson, C. J., and Davies, J. (1991). *Mol. Microbiol.*, **10**, 2499.

54. Kuhstoss, S., Richardson, M. A., and Rao, R. N. (1989). *Gene*, **97**, 143.

55. Khosla, C., Ebert-Khosla, S., and Hopwood, D. A. (1992). *Mol. Microbiol.*, **6**, 3237.

56. Birch, A. W. and Cullum, J. (1985). *J. Gen. Microbiol.*, **131**, 1299.

57. Muth, G., Wohlleben, W., and Pühler, A. (1988). *Mol. Gen. Genet.*, **211**, 424.

58. Muth, G., Nußbaumer, B., Wohlleben, W., and Pühler, A. (1989). *Mol. Gen. Genet.*, **219**, 341.

59. Wohlleben, W., Muth, G., Birr, E., and Pühler, A. (1986). In *Sixth International Symposium on Actinomycetes Biology* (ed. S. Biro and M. Goodfellow), pp. 99–101. Akademiai Kiado, Budapest.

60. Wohlleben, W., Muth, G., and Kalinowski, J. (1993). In *Biotechnology* (ed. H. J.

Rehm, G. Reed, A. Pühler, and P. Stadler), Vol. 2, pp. 455–505. Verlag Chemie, Weinheim.

61. Wohlleben, W., Hartmann, V., Hillemann, D., Krey, K., Muth, G., Nußbaumer, B., and Pelzer, S. (1993). In *Proceedings of the 11th European Meeting on Genetic Transformation*, pp. 171–183. Intercept, Budapest, (in press).
62. MacNeil, T. and Gibbons, P. H. (1986). *Plasmid*, **16**, 182.
63. Hillemann, D., Pühler, A., and Wohlleben, W. (1991). *Nucleic Acids Res.*, **19**, 727.
64. Wohlleben, W. and Pielsticker, A. (1989). In *Dechema biotechnology conferences* (ed. D. Behrens and A. J. Driesel), Vol. 3, pp. 301–5. Verlagsgesellschaft Chemie, Weinheim.
65. Mazodier, P., Petter, R., and Thompson, C. (1989). *J. Bacteriol.*, **171**, 3583.
66. Simon, R., Priefer, U., and Pühler, A. (1983). *Biotechnology*, **1**, 784.
67. Bierman, M., Logan, R., O'Brien, K., Seno, E. T., Rao, N. R., and Schoner, B. E. (1992). *Gene*, **116**, 43.
68. Gough, J. A. and Murray, E. (1983). *J. Mol. Biol.*, **166**, 1.
69. Martin, R. (1987). *Focus*, **9.1**, 11.
70. Chater, K. F. and Bruton, C. J. (1983). *Gene*, **26**, 67.
71. Labes, G., Simon, R., and Wohlleben, W. (1990). *Nucleic Acids Res.*, **18**, 2197.
72. Solenberg, P. J. and Baltz, R. H. (1991). *J. Bacteriol.*, **173**, 1096.
73. Bibb, M. J., Findlay, P. R., and Johnson, M. W. (1984). *Gene*, **30**, 157.

Use of λ–plasmid composite vectors for expression cDNA cloning

TORU MIKI and STUART A. AARONSON

1. Introduction

Plasmid vectors provide straightforward approaches to clone and analyse genes of interest. They are easily manipulated, and a number of versatile plasmid vectors have been developed. Phage vectors have also been developed and successfully applied to cDNA cloning. Use of phage vectors has several advantages. For example, gene expression by λ phage can be much more tightly regulated than with plasmids. Moreover, cell viability is required only when phages are propagating in host cells. Thus, it is believed that λ libraries can be maintained more stably than plasmid libraries. In addition, plaque screening has advantages over colony screening, especially when antibodies are used as probes. High efficiency gene transfer using *in vitro* packaging system is another advantage of λ vectors. In spite of these advantages, analysis of λ clones is usually time-consuming and less efficient. The use of λ–plasmid composite vectors for construction of cDNA libraries combines advantages of both λ and plasmid vectors without any apparent disadvantages. Plasmid libraries can be generated from λ–plasmid libraries by excision of the plasmid carrying cDNA inserts when required. We have developed efficient expression cloning systems using λ–plasmid composite vectors and cloned several genes which are critical to growth control of cells. In this chapter, we describe the use of these systems in detail.

2. Expression cloning using unidirectional cDNA libraries

2.1 The automatic directional cloning (ADC) method

Expression cDNA cloning involves detection of a gene product expressed in bacterial or eukaryotic cells. Probes available include antibodies, binding proteins, or oligonucleotides with binding affinity to the expressed cDNA products. Other approaches involve selection for phenotypic changes in cells

3.4 Amplification of libraries

Amplification of libraries should be performed by preparing plate lysates to avoid recombination between different clones. Up to 5×10^4 plaque-forming units (p.f.u.) of phages can be conveniently plated in a 150 mm dish, although smaller number of plaques are preferred to avoid possible recombination. For transformation of eukaryotic cells, DNA is extracted from phage particles and purified by polyethylene glycol precipitation followed by CsCl equilibrium centrifugation.

4. Analysis of cloned cDNAs

4.1 Plasmid rescue

Following isolation of clones of interest, plasmids are excised from λ DNAs, and the cDNA inserts contained in the plasmids are analysed. When prokaryotic expression cloning is performed, phage DNAs can be prepared from well-isolated single plaques using *Protocol 2* and digested with *Not*I. After phenol/chloroform extraction and precipitation, the digested DNA is ligated and used to transform bacterial cells. In the case of eukaryotic expression cloning, genomic DNA is isolated from G418-selected cultures of interest and plasmid rescue is performed as described in *Protocol 3*.

Protocol 3. Plasmid rescue

1. Digest 1–2 μg of genomic DNA with either of *Not*I, *Xho*I, or *Mlu*I. Extract with phenol/chloroform and then chloroform. Precipitate DNA with ammonium acetate and ethanol.

2. Resuspend DNA in 355 μl of H_2O. Add 50 μl of 10 × ligation buffer (660 mM Tris–HCl pH 7.5, 100 mM $MgCl_2$, 50 mM dithiothreitol, 500 μM ATP), and 5 U of T4 DNA ligase. Incubate overnight at 15 °C. Extract and precipitate the ligated DNA as above, and resuspend in 10 μl of 10 mM Tris–HCl, 1 mM EDTA pH 8.0.

3. Add 1 μl of the ligated DNA to aliquots (100 μl each) of PLK-F' (Stratagene, La Jolla, CA) or DH10B (BRL, Gaithersburg, MD) competent cells. Transform the cells as directed by the manufacturer.

4. After the heat shock, dilute the cells ten-fold with S.O.C. medium (BRL) containing 1 mM IPTG, to induce expression of the *neo* gene driven by the *tac* promoter. Incubate the cultures for 1 h with shaking and plate on NZY hard agar containing ampicillin (100 μg/ml), kanamycin (25 μg/ml), and IPTG (100 μM).

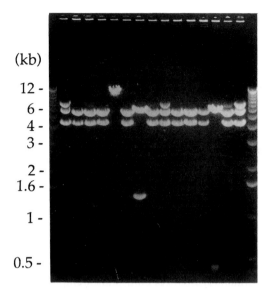

(kb)

12 -
6 -
4 -
3 -
2 -
1.6 -
1 -
0.5 -

Figure 9. *Sal*l digestion of plasmids rescued from a library transformant. Plasmids rescued from a focus induced by transfection of NIH/3T3 cells with a BALB/MK cDNA library in λpCEV27 (14) were digested by *Sal*l and analysed by agarose gel electrophoresis. The 6 kb bands represent plasmid DNA. Clones possessing the 4.2 kb insert scored positive for NIH/3T3 transformation, whereas other clones were negative. The fifth clone from the left with unusual structure may have undergone rearrangement during phagemid propagation, integration, or the plasmid rescue step.

cDNA inserts can be released by *Sal*I digestion. A result of such analysis is shown in *Figure 9*. The efficiency of plasmid rescue varies among transformants. From several to more than 100 colonies can be obtained by this procedure.

4.2 Restriction mapping

From the bacterial transformants, plasmid DNA is extracted and subjected to restriction analysis. In the case of eukaryotic expression cloning, plasmid DNA is also subjected to eukaryotic cell transfection to test whether the clone is responsible for the phenotype of interest. Our protocol for plasmid DNA preparation (1) is described below. We have used this protocol for several years and found it to be useful for a number of purposes including transfection assays. DNA preparations can be stably stored at 4°C.

Protocol 4. Preparation of plasmid DNA by the alkaline lysis/ selective precipitation method

1. Inoculate 10 ml of NZY, L, or super broth containing suitable antibiotics in a Falcon 2070 tube with a single colony. Shake at 37°C overnight.

Table 2. Promoters commonly used in expression systems

Host	Promoter (relative strength when induced)	Inducer	Inhibitor (uninduced basal expression)
E. coli	lac (medium)	IPTG (1 mM in H_2O)	LacI (slightly leaky)
	trp (very strong)	3-indolylacetic acid, indolylacrylic acid	Tryptophan starvation (leaky)
	tac (hybrid trp/lac promoter) (very strong)	IPTG (1mM in H_2O)	LacI (slightly leaky)
	T7 (strong) (ref. 22)	A. IPTG induction of T7 RNA polymerase B. Introduction of T7 polymerase by phage	lacI
	phoA (weak)	Low phosphate medium	High phosphate medium
	araB (weak)	araC plus arabinose	araC
	pNH promoters flanked by attP/attB[a] (strong)	Heat induction for promoter inversion/ IPTG for promoter induction	Growth at 30°C
Mammalian tissue culture cells	Cytomegalovirus immediate early (CMV-IE)	[b]	Constitutive
	Rous sarcoma virus long terminal repeat (RSV-LTR)	[b]	Constitutive
	Metallothionein (MT)	Zn^{2+}, Cd^{2+}	—
	Mouse mammary tumour virus LTR (MMTV-LTR)	Glucocorticoid hormone	
	CMV/lacO[c]	IPTG	LacI

[a] Electroporation must be used for efficient introduction of vector since heat pulse of standard transformation induces the promoters.
[b] Any strong promiscuous *trans*-activator will increase activity (e.g. VP16, BPV-E2, adenovirus E1A).
[c] LacI switch system (see Section 4.5).

contained within vectors that control intracellular RNA levels and thereby affect protein expression. *Table 2* describes some of the more common promoters found in phagemid vectors.

Generalized cloning vectors such as pBluescript allow efficient expression of inserts, sometimes to the level of 10% of the total protein content of the cell. This is often sufficient for preliminary identification and analysis of the protein. *Protocol 2* describes a standard expression procedure which can be used with many general cloning vectors and optimized for particular proteins. For tighter repression of toxic proteins and greater induced transcription levels, *Protocol 3* describes a typical procedure for T7 polymerase induced expression (22).

Peptide tags (e.g. epitope tags, metal chelator tags, protein kinase tags, protease cleavage site tags) are being used more frequently for the purification and labelling of proteins (23–27).

Protocol 2. Small scale protein production using pBluescript for protein identification and Western blot

Materials

- LB (see *Protocol 6* for recipe)
- fresh overnight culture of clone with insert grown in appropriate antibiotic
- fresh overnight culture of host strain harbouring vector without insert in same host strain
- culture tubes, (e.g. 50 ml conical tubes)
- 37°C shaker incubator
- spectrophotometer (set at 600 nm)
- wash buffer: 50 mM Tris pH 8.0, 100 mM NaCl, 1 mM EDTA
- 2×Laemmli buffer (4% SDS, 20% glycerol, 10 mM Tris pH 6.8, 0.003% bromophenol blue)
- 500 mM IPTG, 50 mg/ml ampicillin stock

Method

1. Inoculate 5 ml culture with 0.5 ml of a fresh overnight culture grown in LB media with appropriate antibiotic selection for the plasmid (e.g. for pBluescript clones in XL1-Blue cells, 50 μg ampicillin should be included per ml of LB). A control tube of host bacteria containing the plasmid without insert should be grown in parallel as a negative control. It has been observed that strains vary considerably in their ability to produce different proteins. The choice of *E. coli* host strain is therefore often critical and several strains should be tried to determine the optimal host (28). The choice of media can also be an important factor since it controls the rate of cell growth and protein translation.

2. Grow cultures to mid-to-late log phase (OD_{600} of 0.7 to 0.9 for XL1-Blue cells). Split cultures into two tubes. Induce one of the cultures by adding IPTG to 1 mM final concentration. The optimal cell density for induction varies depending on the clone. Some toxic proteins should be kept repressed as long as possible to prevent premature cell death. The concentration of IPTG required for optimal induction is also plasmid dependent.

3. Allow the cultures to grow under induced conditions for 2–3 h. The optimal length of time for this growth period is again dependent on the clone. Some proteins are sensitive to proteases and longer incubations increase degradation products. Some proteases build up as the cells

Protocol 2. *Continued*

enter stationary phase. Other proteins are very stable and the longer the cells are grown the more protein is produced. Incubation temperature is critical for optimal protein production, folding, and intracellular transport. Sometimes 37°C is optimal whereas other inserts are expressed more efficiently and are able to maintain functional activity at higher or lower temperatures (e.g. room temperature to 42°C for *E. coli*). For instance antibody fragments, when over-produced in *E. coli*, form inclusion bodies from which active protein can not be readily recovered. By growing cells at room temperature, rather than 37°C, expression can be reduced to allow an increase in export of active protein to the periplasm (29).

4. Take a final OD_{600} reading of the cultures. Cool cultures on ice. Remove 1 ml from each culture into Eppendorf tubes. Spin at 6000 *g* for 10 min at 4°C. Wash the pellet once in 100 μl of cold wash buffer. Spin the sample as before and resuspend the pellet in 100 μl Laemmli loading buffer. Boil the samples for 5 min. The samples can be loaded directly on a protein gel (2) or, if they are very viscous, an additional 10 min spin at 12 000 *g* can be done to remove insoluble material. Each sample should be normalized according to the final OD_{600} reading. Usually 1 ml of OD_{600} of 0.5 cells is suitable. Save the original supernatant as well as the Laemmli pellet in case the protein is insoluble or has been secreted into the medium or periplasmic space. The Laemmli gel can be silver stained, Coomassie Blue stained, or analysed by Western blot as desired (2).

Protocol 3. T7 RNA polymerase induction of expression in *E. coli*

Materials

- Vector containing T7 promoter sequence with insert containing appropriate ribosome-binding site for translation initiation. This site is sometimes provided in the vector (e.g. pET-3a (22)) or must be supplied with the insert (pBluescript). Other features required for expression of inserts must also be considered (see *Table 1*)

- Inducible source of T7 RNA polymerase (e.g. BL21(DE3) *E. coli* cells containing IPTG inducible T7 RNA polymerase integrated in the genome, T7 RNA polymerase cloned in M13 phage, or vaccinia virus containing T7 RNA polymerase gene for infection of eukaryotic cells)

- Growth medium with appropriate antibiotic selection for vector containing hosts

Method

Day 1

Inoculate 2–3 ml cultures of growth medium containing the appropriate antibiotic with cells containing the expression vector. A second culture containing the vector without insert should be grown in parallel as a negative control for protein production. Grow overnight at the appropriate temperature.

Day 2

1. Inoculate 5 ml of fresh medium plus antibiotic with 50 μl of the fresh overnight culture. Shake for 2 h at appropriate temperature to mid-log phase.

2. Remove 1 ml of the culture and grow without induction. Induce the remainder of the culture (e.g. by adding IPTG to a final concentration of 1 mM if growing the culture in BL21(DE3) cells, or by infecting with filamentous phage particles encoding T7 polymerase).

3. Remove 1 ml of the culture at 30 min, 1, 2, and 3 h. Immediately centrifuge the samples at 6000 *g* for 1 min at room temperature in a microcentrifuge, and remove supernatants.

4. Resuspend pellets in 100 μl Laemmli buffer and analyse by SDS–PAGE (2). The supernatants from step **3** can also be assayed by SDS–PAGE to determine if the protein has been secreted or otherwise released into the medium (e.g. by cell lysis).

3.3 Phage display

Phage display involves the presentation of fusion proteins on the surface of filamentous phage particles. The main advantage of this system is that it is possible to generate libraries containing more than 10^7 unique members and to isolate target clones from amplified libraries containing 10^9 clones or more. Smith (30) has shown that a peptide/pIII fusion can be displayed on fd phage and is available to specifically bind to antipeptide antibodies. Phage displaying affinity to targeted receptor molecules are selected by a process of affinity selection called 'biopanning' (31). This selection is accomplished by washing or removing unbound phage particles, and recovering the bound fraction. The major coat protein, pVIII, has also been used for presentation of fusion proteins (32). Biopanning of recombinant phage requires only minute amounts of antibody or other ligand. Selected phage are used to infect *E. coli* for further amplification and subjected to additional rounds of enrichment to ensure isolation of phage with beneficial binding activity. DNA inserts within the selected phage are sequenced to determine the peptide structure which give the phage its phenotype.

Phage display libraries have been used to identify peptide epitopes with affinity to specific monoclonal antibodies (31, 33), avidin (34), and conA (35, 36). Phage display has also been used to identify and/or optimize the function of large peptides such as immunoglobulin Fab and single chain Fv fragments (37–41), human growth hormone (41), bovine pancreatic trypsin inhibitors (41), and protein A (R. C Willson, personal communication).

Filamentous phage, phagemid vectors, and excisable lambda phage (44) vectors (see Section 4) have been used for the construction of these libraries. Schematic drawings and the advantages of the major vector-types being used are described in *Figure 2*. Methods for insert construction and biopanning

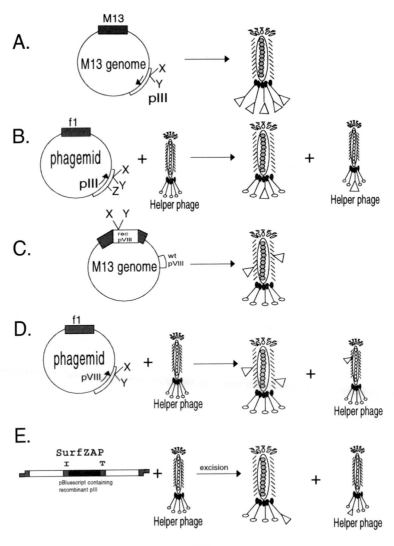

vary depending on the experiment. The information cited above should be consulted for detailed protocols.

3.4 Polycos vectors

Polycos vectors (43) are filamentous phage or phagemids which contain a lambda packaging recognition sequence, the *cos* site, and can be introduced into *E.coli* cells as concatemers (polymers) through the use of *in vitro* lambda packaging extracts (see *Protocol 4*). In order to satisfy the requirements for lambda packaging, the polycos vectors are digested with restriction enzymes to form linear molecules and ligated at high DNA concentrations to enhance the production of concatemers (see *Figure 3*). Proteins within the lambda packaging extract select a *cos* site at random within the concatemer and then scan down the DNA molecule until a second *cos* site in the same orientation is encountered. This second *cos* site will only be detected after the lambda phage head is approximately full, i.e. greater than 38 kb, but less than 51 kb.

Figure 2. (A) p3 vectors. Fragments are cloned directly into the pIII gene of the filament-ous phage genome. Cloning sites are positioned at the amino-terminus of the mature spliced protein between sites X and Y (after the leader peptide cleavage site). pIII vectors allow the detection of low affinity binding activity by having all five copies of fusion pIII molecules thereby increasing the avidity of binding. If the fusion protein results in a pIII protein which can no longer recognize the receptors on the pili, the resulting phage will not be infectious and is difficult to amplify and identify. (B) p3+3 vectors. Fragments are cloned into a pIII gene incorporated in a phagemid vector while using helper phage to supply all other proteins required for packaging the single-stranded phagemid DNA (as in single-strand rescue Section 2.5). An inducible promoter, such as the phoA promoter, can be used to regulate the number of recombinant pIII molecules on the surface of the phagemid. As there is a mixture of wild-type and recombinant pIII molecules on the surface of the phagemids, infectivity is greatly improved. Cloning sites are positioned at the amino-terminus as in pIII vectors. The entire amino-terminal domain (amino acids 1–198, between sites X and Z) can be replaced by the insert peptide when larger fusions, such as Fab fragments, are being cloned. (C) p88 vectors. A synthetic pVIII gene is inserted into the multicloning site of an M13 vector. The wobble base of each codon is changed to prevent homologous recombination between the wild-type and recombinant genes while maintaining the native amino acid sequence. Inserts are cloned into the amino-terminus of the spliced recombinant pVIII protein. Both the recombinant and wild-type pVIII are expressed in this vector, allowing regulation of the number of recombinant pVIII proteins. Since there are at least 2600 pVIII proteins on the surface of each phage, there is the potential for high avidity. It is usually not possible to present 100% recombinant pVIII protein on the phage since inserts into pVIII prevent proper phage particle formation. (D) p8+8 vectors. As in the p3+3 vector system, the recombinant pVIII gene is present on a phagemid, and helper phage supply all proteins required for packaging of the phagemid. As in the p88 vectors, the cloning sites are positioned to fuse the insert on to the amino-terminus of the mature spliced pVIII protein. (E) Excisable vectors. The lambda Surf-ZAP™ vector allows high efficiency cloning of immunoglobulin Fab fragments. These immunoglobulin libraries are excised and then screened by phage display in a similar manner to the p3+3 expression system. $\triangle$ = protein or peptide of interest.)

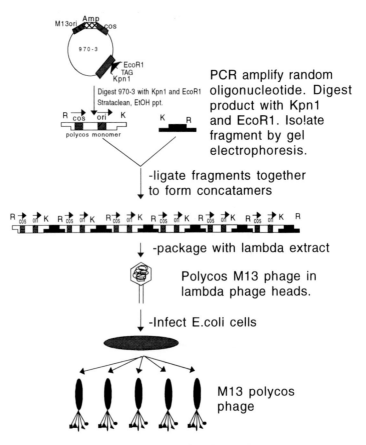

PCR amplify random oligonucleotide. Digest product with Kpn1 and EcoR1. Isolate fragment by gel electrophoresis.

-ligate fragments together to form concatamers

-package with lambda extract

Polycos M13 phage in lambda phage heads.

-Infect E.coli cells

M13 polycos phage

Figure 3. Polycos vector packaging process. See *Protocol 4*.

For a 7.8 kb filamentous phage 'polycos' vector, five or six copies would be packaged into each lambda phage head as a head-to-tail concatemer.

Release of M13 monomers from the concatemer occurs by a system analogous to helper phage excision of the phagemid pBluescript from lambda ZAP (see Section 4.1). Upon entering the cell, M13 pII protein is expressed. This catalyses the formation of a nick at the M13 origin of replication, promoting rolling circle replication. Following displacement of the parental strand and synthesis of a new DNA strand, pII protein generates a second nick at the terminator site within the next M13 origin, and then seals the ends of the displaced strands to generate circular, single-stranded monomeric M13 molecules. By utilizing this system, newly synthesized monomers are generated from the parent molecule. Normal replication of the monomers continues and packaged phagemids or filamentous phage particles are extruded through the membrane. When cloning sites within pIII are used, the recombinant protein is expressed on the surface of the phage particles.

Protocol 4. Polycos ligation and packaging

Materials

- phagemid or filamentous phage vector containing lambda *cos* site
- ligase and 10 × ligase buffer (500 mM Tris–
  HCl pH 7.5, 7 mM MgCl$_2$, 1 mM DTT, 10 mM
  ATP)
- double-stranded DNA insert
- GigapackII *in vitro* lambda packaging extracts

- restriction enzyme and buffer
- phenol/chloroform (1:1, v/v)
- ethanol
- 3 M sodium acetate (NaAc) pH 5.2

Method

1. Linearize 10 μg of vector DNA with 30 U of restriction enzyme for 1 h at
   the appropriate temperature.

2. Add an equal volume of phenol/chloroform. Mix well. Centrifuge at
   12 000 *g* for 5 min. Transfer the upper phase to a fresh Eppendorf tube,
   add 0.1 volume of 3 M NaAc pH 5.2, 2 volumes of 100% ethanol, and
   mix well. Spin at 12 000 *g* for 15 min. Remove the ethanol supernatant.
   Spin briefly and remove any remaining ethanol, being careful not to
   disturb the DNA pellet.

3. Resuspend DNA at a concentration of 1 μg/μl in water.

4. Mix 2 μg of digested DNA with 0.5 μl of 10 × ligation buffer, 2.5 μl of
   water, and 0.5 μl of ligase (final volume of 5 μl). Double-stranded insert
   DNA with compatible restriction overhangs are added to tubes prepared in parallel at molar ratios of 0.5:1, 1:1, 5:1.

5. Ligate overnight at 4°C.

6. Package the concatemerized DNA with packaging extract according to
   manufacturer's protocol.

7. Plate for colonies or plaques, as appropriate for the vector, to determine the library size.

3.5 PCR cloning with phagemid vectors

Several specialized vectors have been developed which facilitate the cloning
of PCR products. The most common method of PCR fragment cloning
involves the engineering of convenient restriction sites on the flanking ends of
the PCR primers, allowing cleavage of the amplification product with enzymes appropriate to the cloning scheme. A second commonly used method
of cloning PCR fragments is to simply clone the unprocessed fragment into a
blunt cloning site in the vector. 'T-A' cloning vectors (45) take advantage of
the 3'-dAMP terminal transferase activity of Taq polymerase by providing a
cloning site containing 3'-dTMP overhangs. 'Ligation-independent PCR clon-

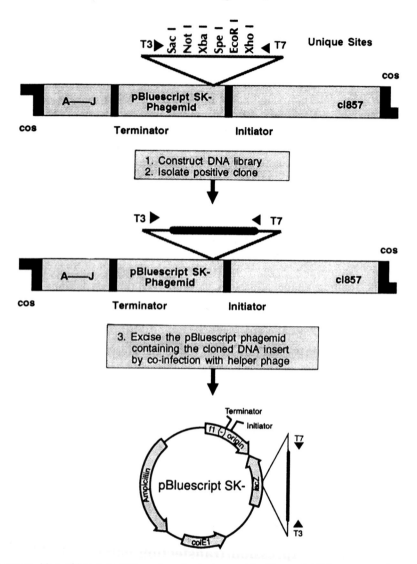

Figure 4. Map of Lambda ZAP and excised phagemid, pBluescript SK−.

cloning capacity of approximately 10 kb. Inserts are expressed in prokaryotic cells using the *lac* promoter and in eukaryotic cells using the cytomegalovirus (CMV) immediate early promoter. The alpha-complementing portion of the *lacZ* gene allows prokaryotic blue/white colour detection of clones that have inserts and enables the expression of fusion proteins. Like lambda ZAP, ZAP Express allows inserts to be excised with the use of helper M13 phage (*Protocol 6*). The excised phagemid is the kanamycin/geneticin resistant pBK-CMV.

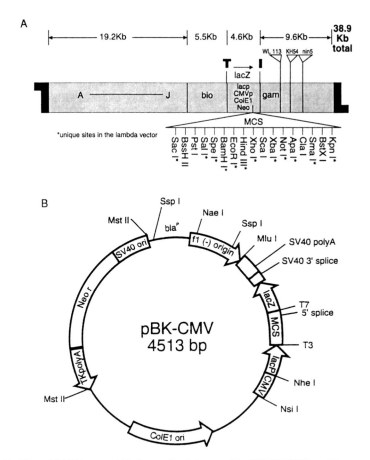

Figure 5. Map of ZAP Express (A), the excised phagemid, pBK-CMV (B), and the expression cassette within these vectors (C).

Directional cDNA libraries constructed using *Xho*I and *Eco*RI restriction enzymes (49) can be ligated into ZAP Express in either the sense or antisense orientation using *Sal*I or *Xho*I sites located on either side of the *Eco*RI site. This is useful for constructing subtraction hybridization libraries and for generating subtracted tissue-specific radioactive probes (51).

Screening of libraries in eukaryotic cells can be performed either by transient or stable transfection of cells (52) using phagemid DNA derived by mass excision of the library (see *Protocol 7*).

4.3 Single clone excision

The *in vivo* excision of individual lambda clones is best performed with filamentous phage containing an amber mutation in the gene II protein. The

4.4 Mass excision

Although the largest libraries are most simply constructed using lambda phage vectors, there are sometimes advantages in having the entire library in phagemid vectors. This is particularly true when constructing subtraction or phage display libraries. It is possible to convert an entire lambda ZAP library into a phagemid library, but excision times should be kept to a minimum in order to minimize changes in the representation of clones within the library. We have previously compared the representation of clones following plate amplification of lambda phage to that of liquid phagemid excision (15). These studies demonstrate that liquid excision times of less than three hours are comparable to lambda plate amplification. The excised clones can be further amplified as colonies by infecting cells that are non-permissive for filamentous phage replication. The resulting cultures can be used for maintaining the library, and for preparing increased amounts of phagemid DNA which can then be used for hybrid selection against other libraries (48).

Protocol 7. Mass excision

- materials as for *Protocol 6.*
- lambda ZAP-based library for which both

the primary library size and the titre of the amplified stock are known.

Method

1. In a 50 ml conical tube, combine 1 ml of XL1-Blue MRF′ cells (OD_{600} of 5.0) with 4×10^{10} ExAssist helper phage. Add enough lambda ZAP bacteriophage such that 100-fold excess of the primary library size is used. For example, if the primary library contained 10^6 members, use 10^8 phage particles from the amplified library stock. Incubate for 15 min at 37 °C.

2. Add 20 ml of LB medium and incubate with gentle agitation for 2 to 3 h. *Note*: incubation times in excess of 3 h can significantly alter library representation (15). Spin down cell debris. The supernatant contains the excised library as filamentous phage particles.

3. Titre the supernatant as in *Protocol 6* steps 5–6 for antibiotic resistant c.f.u. with XLOLR cells on LB/ampicillin plates (LB/kanamycin plates for ZAP Express) to determine the number of excised phagemids.

Amplification of excised libraries

4. Mix the recovered supernatant from step **2** with enough XLOLR cells so that there are at least ten cells per c.f.u. (assume 8×10^8 cells per ml of OD_{600} of 1.0 cells) and incubate at 37 °C for 15 min.

5a. Plate amplification. Add an equal volume of LB, incubate 30 min at 37°C, and plate on LB/ampicillin plates (LB/kanamycin plates for ZAP Express). Use 10^4 c.f.u. per plate. Incubate the plates overnight at 37°C for colony amplification.

OR

5b. Liquid amplification. Add 20 ml of LB medium containing 50 μg/ml antibiotic and allow the cells to grow for 6 h at 37°C.

6. The antibiotic-resistant *E. coli* cells recovered from step **5a** or **5b** can be used for the isolation of double-stranded DNA by standard plasmid preparation procedures. Single-stranded material can be recovered by infecting the cultures with helper phage and utilizing single-stranded rescue procedures as described in *Protocol 1*. These DNA pools can be used for hybrid selection against other libraries, RNA, or DNA samples.

4.5 New applications

4.5.1 Phage and phagemids in genetic toxicology

The process of *in vivo* phagemid excision from lambda vectors has recently been utilized in a lambda phage shuttle vector system for genotoxicity testing in transgenic animals (Big Blue™) (53). Through microinjection of single cell embryos, an excisable lambda phage (Lambda LIZ) was stably integrated into both mouse and rat chromosomes. Mutations occurring within the genes contained within the lambda phage (e.g. *lacI*, or *lacZ*) can be scored by directly packaging the phage integrated in the genomic DNA with *in vitro* lambda packaging extracts and plating with *E. coli* cells. Mutations within the target gene can be detected by the phenotype conferred upon the phage plaque formed by the replicating lambda phage. The partial f1 origins within this lambda vector permit excision of the target gene to generate a phagemid template for sequence characterization. This system presents the first opportunity to directly measure tissue-specific mutation frequencies and to investigate the underlying mechanisms of mutagenesis and carcinogenesis through sequence analysis.

4.5.2 LacI switch phagemid

Phagemid vector systems are also improving the ability to regulate genes in eukaryotic cells. In addition to T7 inducible systems (22), extensive work by a number of research groups has led to the adaptation of bacterial *lacI* regulatory systems to eukaryotic cells (54). This approach has the advantage that the lac control system has no function within the eukaryotic cell and therefore, regulation of a cloned gene should be solely dependent upon the exogenous inducer, IPTG. In order for this system to function within mammalian cells, the *lac* repressor gene is expressed via a eukaryotic promoter. The target gene, whose expression is to be regulated, is positioned downstream from an 18 base pair *lac* operator sequence. LacI repressor protein binds to the *lac* operator to inhibit transcription of the target gene by the eukaryotic promoter. Target gene expression is not induced until the galactoside inducer is added into the culture. This approach and other similar methods will be valuable in the study of temporal responsive genes controlling cell development and transformation.

5. Future

In light of the recent technological advances which allow rapid amplification of minute quantities of DNA (PCR) (55), it is conceivable that in the future the need for DNA vectors will be eliminated altogether. One can imagine scenarios in which PCR-amplified fragments are used to transform appropriately engineered *E. coli* strains directly rather than first subcloning the fragment into a plasmid vector. 'Insert-plus' positive clones could be screened as chromosomal inserts, and functional analysis, such as induction of protein expression, may be performed directly from the chromosome. Since one can obtain a substantial quantity of specific DNA in a few hours using PCR amplification, PCR has substantially reduced the need for plasmid-based gene amplification and analysis. PCR-based techniques have been developed which allow genomic DNA to be sequenced directly without subcloning (56, 57). More recently, efficient *in vitro* transcription/translation systems have been developed in which the PCR product is expressed directly in the reaction tube (58). Such systems completely eliminate the common *in vivo* expression problem, namely the toxicity of the cloned protein to the bacterial host. Theoretically, adaptation of this system to a continuous cell-free translation system (59) should allow preparative 'Coomassie gel' quantities of protein from *in vitro* expression systems using unpurified PCR products. The rapidly growing number of thermostable enzymes that have been characterized and cloned from the genomes of a variety of thermophilic bacteria will result in more stable *in vitro* enzymatic systems, eliminate the need for refrigeration, and possibly further hasten the trend from vector-based systems. Nevertheless, DNA vectors are at present the most versatile system. PCR amplification of large DNA fragments is not efficient. Many techniques will continue to benefit from the use of DNA vectors: the amplification of large DNA fragments, large scale DNA and protein preparations, and the construction, screening, analysis, and maintenance of cloned genes and libraries.

References

1. Messing, J. (1983). In *Methods in enzymology* (ed. R. Wu, L. Grossman, and K. Moldave), Vol. 101, pp. 20–78. Academic Press, London.
2. Sambrook, J., Fritsch, E. F., and Maniatis, T. (ed.) (1989). *Molecular cloning, a laboratory manual* (2nd edn), pp. 15. 51–15. 80. Cold Spring Harbor Laboratory Press, NY.
3. Rasched, I. and Oberer, E. (1986). *Microbiol. Rev.*, **50**, 401.
4. Kornberg, A. (ed.) (1980). *DNA Replication*. W. H. Freeman & Co., San Francisco, CA.
5. Messing, J., Gronenborn, B., Muller-Hill, B., and Hofschneider, P. (1977). *Proc. Natl. Acad. Sci. USA*, **74**, 3642.
6. Messing, J. and Vieira, J. (1982). *Gene*, **19**, 269.
7. Yanisch-Perron, C., Vieira, J., and Messing, J. (1985). *Gene*, **33**, 103.
8. Zinder, N. D. and Boeke, J. D. (1982). *Gene*, **19**, 1.
9. Bolivar, F., Rodriguez, R. L., Greene, P. J., Betlach, M. C., Heynecker, H. L., and Boyer, H. W. (1977). *Gene*, **2**, 95.
10. Dotto, G. P., Horiuchi, K., and Zinder, N. D. (1984). *J. Mol. Biol.*, **172**, 507.
11. Dente, L., Cesareni, G., and Cortese, R. (1983). *Nucleic Acids Res.*, **11**, 1645.
12. Smith, G. P. (1988). In *Vectors, a survey of molecular cloning vectors and their uses* (ed. R. L. Rodriguez and D. T. Denhardt), pp. 61–84. Butterworth Publishers, Stoneham, MA.
13. Katayama, C. (1990). *Strategies*, **3**, 56.
14. Russel, M., Kidd, S., and Kelley, M. R. (1986). *Gene*, **45**, 333.
15. Hay, B. and Short, J. M. (1992). *Strategies*, **5**, 16.
16. Short, J. M., and Sorge, J. A. (1992). In *Methods in enzymology* (ed. R. Wu.), Vol. 216, pp. 495–516. Academic Press, London.
17. Alting-Mees, M. A., Sorge, J. A., and Short, J. M. (1992). In *Methods in enzymology* (ed. R. Wu.), Vol. 216, pp. 483–95. Academic Press, London.
18. Brown, T. A. (ed.) (1991). *Molecular Biology LabFax*. BIOS Scientific Publishers Limited, Oxford.
19. Rodriguez, R. L. and Denhardt, D. T. (ed.) (1988). *Vectors, a survey of molecular cloning vectors and their uses*. Butterworth Publishers, Stoneham, MA.
20. Goeddel, D. V. (ed.) (1990). *Methods in enzymology*, Vol. 185. Academic Press, London.
21. Davies, J. and Rosenberg, M. (ed.) (1992). *Current opinion in biotechnology*. Vol. 3, pp. 454–594. Current Biology, London.
22. Studier, F. W., Rosenberg, A. H., Dunn, J. J., and Dubendorff, J. W. (1990). In *Methods in enzymology* (ed. D.V. Goeddel), Vol. 185, pp. 60–8. Academic Press, London.
23. Hopp, T. P. (1986). *J. Immunol. Methods*, **88**, 1.
24. Carter, P. (1990). In *ACS Symposium Series, Protein Purification*. (ed. M. R. Ladisch, R. C. Willson, C. C. Painton, and S. E. Builder), pp. 181–93. American Chemical Society, Washington.
25. Smith, M. C., Cook, J. A., Furman, T. C., Gesellchen, P. D., Smith, D. P., and Hsiung, H. (1990). In *ACS Symposium Series, Protein Purification* (ed. M. R. Ladisch, R. C. Willson, C. C. Painton, and S. E. Builder), pp. 168–80. American Chemical Society, Washington.
26. Persson, M., Bergstrand, M. G., Bulow, L., and Mosbach, K. (1988). *Anal. Biochem.*, **172**, 330.

27. Blanar, M. A. and Rutter, W. J. (1992). *Science*, **256**, 1014.
28. Hatt, J., Callahan, M., and Greener, A. (1992). *Strategies*, **5**, 2.
29. Cabilly, S. (1989). *Gene*, **85**, 553.
30. Smith G. P. (1985). *Science*, **228**, 1315.
31. Parmley, S. F. and Smith G. P. (1988). *Gene*, **73**, 305.
32. Felici, F., Castagnoli, L., Musacchio, A., Jappelli, R., and Cesareni, G. (1991). *J. Mol. Biol.*, **222**, 301.
33. Cwirla, S. E., Peters, E. A., Barrett, R. W., and Dower, W. J. K. (1990). *Proc. Natl. Acad. Sci. USA*, **87**, 6378.
34. Devlin, J. J., Panganiban, L. C., and Devlin, P. E. (1990). *Science*, **249**, 404.
35. Scott, J. K., Loganathan, D., Easley, R. B., Gong, X., and Goldstein, I. J. (1992). *Proc. Natl. Acad. Sci. USA*, **89**, 5398.
36. Oldenburg, K. R., Loganathan, D., Goldstein, I. J., Schultz, P. G., and Gallop, M. A. (1992). *Proc. Natl. Acad. Sci. USA*, **89**, 5393.
37. Kang, A. S., Barbas, C. F., Janda, K. D., Benkovic, S. J., and Lerner, R. A. (1991). *Proc. Natl. Acad. Sci. USA*, **88**, 4363.
38. McCafferty, J., Griffiths, A. D., Winter, G., and Chiswell, D. J. (1990). *Nature (London)*, **348**, 552.
39. Gram, H., Marconi, L.-A., Barbas, C. F., Collet, T. A., Lerner, R. A., and Kang, A. S. (1992). *Proc. Natl. Acad. Sci. USA*, **89**, 3576.
40. Embleton, M. J., Gorochov, G., Jones, P. T., and Winter, G. (1992). *Nucleic Acids Res.*, **20**, 3831.
41. Bass, S., Greene, R., and Wells, J. A. (1990). *Proteins*, **8**, 309.
42. Roberts, B. L., Markland, W., Ley, A. C., Kent, R. B., White, D. W., Guterman, S. K., and Ladner, R. C. (1992). *Proc. Natl. Acad. Sci. USA*, **89**, 2429.
43. Alting-Mees, M. A. and Short, J. M. (1993). (Submitted for publication).
44. Hogrefe, H. H., Mullinax, R. L., Lovejoy, A. E., Hay, B. N., and Sorge, J. A. (1993). *Gene*, **128**, 119.
45. Mead, D. A., Pey, N. K., Hernstadt, C., Mercil, R. A., Smith, L. M. (1991). *Bio/Technology*, **9**, 657.
46. Aslandis, C. and de Jong, P. J. (1990). *Nucleic Acids Res.*, **18**, 6069.
47. Bauer, J., Deely, D., Braman, J., Viola, J., and Weiner, M. (1992). *Strategies*, **5**, 62.
48. Short, J. M., Fernandez, J. M., Sorge, J., and Huse, W. D. (1988). *Nucleic Acids Res.*, **16**, 7583.
49. Alting-Mees, M. A., Sorge, J. A., and Short, J. M. (1992). In *Methods in enzymology* (ed. R. Wu), Vol. 216, pp. 483–95. Academic Press, London.
50. Huse, W. D. and Hansen, C. (1988). *Strategies*, **1**, 1.
51. Schweinfest, C. W., Henderson, K. W., Gu, J.-R., Kottaridis, S. D., Besbeas, S., Panotopoulou, E., and Papas, T. S. (1990). *Gene Anal. Tech. Appl.*, **7**, 64.
52. Seed, B. and Aruffo, A. (1987). *Proc. Natl. Acad. Sci. USA*, **87**, 3365.
53. Kohler, S. W., Provost, G. S., Fieck, A., Kretz, P. L., Bullock, W. O., Sorge, J. A., Putman, D. L., and Short, J. M. (1991). *Proc. Natl. Acad. Sci. USA*, **88**, 7958.
54. DuCoeur, L., Wyborski, D. L., and Short, J. M. (1992). *Strategies* **5**, 70.
55. Innis, M. A., Gelfand, D. H., Sninsky, J. J., and White, T. J. (1990). *PCR protocols*. Academic Press, San Diego.
56. Kretz, K., Hedden, V., and Callen, W. (1992). *Strategies*, **5**, 32.

57. Wong, C., Dowling, C. E., Saiki, R. K., Higuchi, R. G., Erlich, H. A., and Kazazian, H. H. (1987). *Nature*, **330**, 384.
58. Kain, K. C., Orlandi, P. A., and Lanar, D. E. (1991). *Biotechniques*, **10**, 366.
59. Spirin, A. S., Baranov, V. I., Ryabova, L. A., Ovodov, S. Y., and Alakhov, Y. B. (1988). *Science*, **242**, 1162.
60. Kozak, M. (1986). *Cell*, **44**, 283.
61. Alting-Mees, M., Huener, P., Ardourel, D., Sorge, J. A., and Short, J. M. (1992). *Strategies*, **5**, 58.

9

Extrachromosomal cloning vectors of *Saccharomyces cerevisiae*

ROBERT S. SIKORSKI

1. Introduction

The baker's yeast *Saccharomyces cerevisiae* has become the experimental organism of choice for a diverse group of investigators. Sophisticated genetic techniques can be used to examine such fundamental cellular processes as cell cycle control, protein processing, gene expression, or receptor signalling. The transfer and expression of heterologous genes in yeast has provided evidence that many of these processes share mechanisms conserved throughout evolution.

The complex manipulations of genes now possible in yeast stem from the development of efficient DNA-mediated transformation techniques involving both integrative and episomal cloning vectors. Yeast episomal cloning vectors are analogous to bacterial plasmids in both their overall utility and impact on genetic studies. In fact, episomes which replicate autonomously in yeast can often be shuttled between yeast and bacteria without modification. The yeast shuttle vectors that are currently used are the product of a continuous attempt to incorporate DNA elements offering improved cloning efficiency and versatility.

This chapter reviews both the basic and specialized DNA elements required to build the various episomal cloning vehicles of *S. cerevisiae*. The molecular components and structural organization of the common types of vectors are described first, followed by a description of functional elements which can be used to customize vectors for individual needs. Protocols are also described for introducing episomal DNA into *S. cerevisiae*. In addition to allowing the reader to manipulate most existing yeast extrachromosomal vectors, it is hoped that the information in this chapter will be of use to those who wish to construct the next generation of yeast cloning vectors.

2. Basic cloning vectors of S. cerevisiae

2.1 Functional elements

With the exception of a bacterial plasmid backbone, the elements used for yeast vector construction and modification (*ARS*, *CEN*, selectable genes, etc.) are contained within rather small DNA fragments (100–2500 bp). Practically this means that the polymerase chain reaction can be used to rapidly isolate the DNA elements needed for vector construction directly from total yeast genomic DNA. Alternatively, the appropriate elements can be obtained from fragments of existing yeast shuttle vectors.

2.1.1 Bacterial plasmid backbone

The ability to propagate yeast shuttle vectors in both *S. cerevisiae* and *E. coli* depends on the presence of yeast-specific (see below) and bacterial-specific replication elements. The prototypic yeast shuttle vectors incorporated the backbone of the *E. coli* vector pBR322 and consequently contained a bacterial origin of DNA replication and one or two functioning selectable bacterial genes (*Ap*[r] and *Tc*[r]). Advances in bacterial vector technology have led to the creation of second and third generation derivatives of pBR322 such as pUC (1) and pBluescript (2) which are better suited for shuttle vector construction. These newer plasmid vectors are smaller in size and possess a greater number of unique restriction enzyme recognition sites for cloning. In addition, both pUC and pBluescript replicate to copy numbers 10–20-fold greater than pBR322 and therefore produce significantly higher DNA yields for purification and subsequent manipulation.

2.1.2 Autonomously replicating sequence

Maintenance of an extrachromosomal cloning vector in *S. cerevisiae* requires the addition of a yeast origin of DNA replication or autonomously replicating sequence (*ARS*). Several DNA replication origins (for example *ARS1*, chromosome III *ARS*, and *ARSH4*) have been isolated from the yeast genome by virtue of their modular nature and their ability to promote the high frequency transformation of circular DNA molecules. Deletion mutagenesis studies have shown that full replication function can be achieved using small *ARS*-containing DNA fragments (3, 4). For *ARS1*, only 193 bp DNA containing three conserved structural subdomains is sufficient (5). For yeast vector construction, there does not appear to be any advantage of one *ARS* over another.

2.1.3 Yeast centromere

An episomal vector in *S. cerevisiae* can be made to segregate with a higher fidelity and lower copy number by adding a yeast centromere (CEN) sequence. *ARS*-bearing plasmids segregate with a bias during mitosis so that over several generations mother cells accumulate plasmids to a high copy number while daughter cells suffer a high rate of plasmid loss. Centromere elements are sequences of genomic DNA which link the endogenous yeast chromosomes to the mitotic spindle apparatus during mitosis and ensure equal partitioning of genetic material. Molecular clones of all *S. cerevisiae* centromeres have been isolated (6), and like *ARS* elements, *CEN*s have been shown to function in a modular fashion. Mutagenesis studies have shown that a 126 bp segment of chromosome III contains all of the *cis* elements required for providing episome stability and for maintaining a low episome copy number (7).

2.1.4 2μ circle DNA replication origin

The *cis* and *trans* elements which control amplification and distribution of the 2μ circle plasmid offer an alternative to cloned chromosomal *ARS* and *CEN* elements as a means for stabilizing the segregation of yeast shuttle vectors. Most strains of *S. cerevisiae* contain 50–100 copies of a 6.3 kb non-essential episome of unknown function called the 2μ circle. Genetic analysis has revealed that the *REP1* and *REP2* genes of 2μ circle encode proteins which act together on the *cis* sequence, *REP3*, to promote equal partitioning of replicated 2μ circle DNA during mitosis (8, 9). Yeast shuttle vectors need only contain the *REP3* sequence and an *ARS* (one exists naturally in 2μ and can be included in a restriction fragment with *REP3* gene) for full replication and partitioning functions, since both *REP1* and *REP2* proteins can be provided in *trans* by ubiquitous, endogenous 2μ plasmids.

2.1.5 Selectable yeast gene

Because of the relatively frequent mitotic missegregation of even the most stable yeast shuttle vectors, long-term maintenance of extrachromosomal vectors necessitates a method for constant selection of plasmid-containing progeny. In contrast to bacterial cloning vectors which usually incorporate antibiotic resistance loci for selection of transformed clones, yeast cloning vectors carry genes which complement genetic defects within conditional biosynthetic pathways. Listed below are the most commonly used genes for prototroph selection in *S. cerevisiae* together with the growth medium supplement necessary for propagation of each cognate yeast mutant.

(a) *HIS3* encodes imidazoleglycerolphosphate dehydrogenase and is required for the biosynthesis of the amino acid histidine (10).

(b) *TRP1* encodes the enzyme phosphoribosyl-anthranilate synthesis and is required for the biosynthesis of the amino acid tryptophan (11). The tight physical linkage of *TRP1* to *ARS1* has been exploited to create the most minimal of YRp vectors by simply circularizing a genomic restriction fragment containing the two elements (12).

(c) *LEU2* encodes the enzyme β-isopropylmalate dehydrogenase and is required for the biosynthesis of the amino acid leucine (13). A deletion allele termed *LEU2D*, made by removing a portion of the *LEU2* promoter region, can be used to amplify yeast plasmids to high copy number because the LEU$^+$ phenotype of *LEU2D* is only expressed when cells accumulate approximately 100 copies of the partially inactivated gene (14).

(d) *URA3* encodes the enzyme orotidine-5′-phosphate decarboxylase and is required for the biosynthesis of the nucleotide uracil (15). An additional possibility when using a *URA3*-based episome is the ability to select *against* the growth of (to counterselect) those cells which harbour this particular plasmid. Counterselection with *URA3* is accomplished using the drug 5-fluoroorotic acid (5-FOA) which is modified into a lethal product by the *URA3* decarboxylase of the URA3$^+$ cells but remains inert in ura3$^-$ cells (16).

(e) *LYS2* encodes the enzyme α-aminoadipate reductase and is required for the biosynthesis of the amino acid lysine (17). As with *URA3*, both selection and counterselection techniques can be used with *LYS2*-containing shuttle vectors. Counterselection is based on the observation that growth of LYS2$^+$ yeast strains is prevented by the drug α-aminoadipate (18).

2.1.6 Cloning sites

Unique restriction endonuclease recognition sites are used to clone DNA into the backbone of yeast shuttle vectors. In contrast to the prototypic yeast cloning vectors based on pBR322 in which existing natural restriction sites were used for cloning, newer cloning vectors often contain an impressive array of synthetic cloning sites grouped within a polylinker. The polylinker regions of pUC (various cloning sites) and pBluescript (*Kpn*I, *Apa*I, *Xho*I, *Sal*I, *Cla*I, *Hin*dIII, *Eco*RV, *Eco*RI, *Pst*I, *Sma*I, *Bam*HI, *Spe*I, *Xba*I, *Eag*I, *Not*I, *Bst*XI, *Sac*II, *Sac*I) have been incorporated into several yeast vectors (*Table 1*).

2.1.7 Telomere sequences

The maintenance of linear plasmids in yeast necessitates the function of another DNA element, the telomere, a specialized structure which serves to protect DNA termini from degradation and to allow completion of DNA replication. Tandem oligonucleotide repeats from the termini of natural yeast chromosomes have been cloned and shown to be sufficient for the formation of *de novo* telomeres within *S. cerevisiae* (19). Thus telomeres can be incorporated into a linear cloning vector by simply transforming a linearized plasmid construct flanked by cloned telomeric repeats. Telomeric repeats can be ligated to very large fragments of heterologous passenger DNA to produce yeast artificial chromosomes (see Section 3.2.6).

2.2 Basic types of yeast extrachromosomal cloning vectors

A schematic diagram depicting the molecular anatomy of yeast extrachromosomal cloning vectors is shown in *Figure 1*. A comprehensive list of currently available shuttle vectors and their sources has been compiled in an excellent review by Rose and Broach (20). Other useful shuttle vector derivatives have also been reported (21–23).

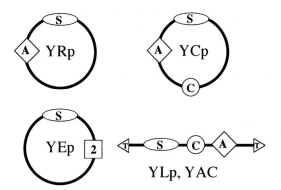

Figure 1. The extrachromosomal cloning vectors of *S. cerevisiae*. Shown are schematic representations of the molecular anatomy and overall structural arrangement of the four basic vectors which can replicate autonomously in this yeast. Symbols represent the pertinent yeast functional elements used in the construction of each vector (A = ARS, S = selectable yeast gene, C = centromere, 2 = 2μ origin of DNA replication, and T = telomere). The bacterial plasmid backbone is shown as a *bold line*.

2.2.1 YRp

Yeast replicating plasmids (YRp) are the most primitive episomal cloning vectors of *S. cerevisiae*, containing only plasmid backbone, yeast *ARS*, and yeast selectable gene. YRp vectors are maintained at high copy number in yeast (approximately 100 per cell) but are so unstable that even under nutrient selection a YRp vector may be present in only 20% of the yeast population. Although YRp vectors may have very specific applications, such as the functional cloning of centromeres (8) or the *in vivo* over-production of DNA elements, their main use is simply to provide a starting point for building other vector derivatives.

2.2.2 YCp

Yeast centromere plasmids (YCp) are one such derivative of YRp vectors made by addition of a *CEN* sequence. YCp vectors are maintained at low copy number in yeast (one or two copies per cell) and are stable enough that YCp episomes can be detected in approximately 90% of cells from cultures grown without selection over ten generations. YCp vectors serve as the workhorses in most yeast cloning projects, and they are particularly valuable for manipulating genes which may be lethal to yeast if present in high copy. Yeast genes are often cloned through expression of their function by transforming clone banks constructed in YCp vectors.

2.2.3 YEp

The so-called yeast episomal plasmids (YEp) contain a bacterial plasmid backbone, a selectable yeast gene, and the *REP3* gene with its flanking sequences from the 2μ circle. YEp vectors are maintained at about 10–40 copies per cell depending on the size and nature of other DNA elements within the plasmid. Incorporation of the 2μ circle partitioning system into the YEp achieves a high degree of plasmid stability; the overall rate of mitotic missegregation of YEp vectors approaches that of a YCp. Their high copy number and relative stability makes YEp vectors particularly well suited to form the backbone of yeast expression vectors. YEp vectors are also useful when over-expression of the genomic clone of a particular gene is needed or is to be directly selected. Note that the replication properties of a YEp demand that the yeast host also harbours a wild-type 2μ circle as a source of essential *trans*-acting gene products.

2.2.4 YLp

Yeast linear plasmids (YLp) are essentially linear versions of YCp vectors which have telomere sequences at their two termini in addition to *CEN, ARS,* and selectable elements. *CEN* function prevents the copy number of YLp vectors from rising beyond one or two per cell. The mitotic stability of any particular YLp is dependent on the total size of the construct, with larger 'artificial chromosomes' missegregating at mitosis less frequently than smaller ones (24). YLp vectors can be propagated in *E. coli* in a circular form for easy manipulation during cloning steps and linearized prior to introduction into yeast. Although there is little role for small YLp vectors, YLp derivatives termed YACs are frequently used for cloning and modifying extremely large fragments of passenger DNA (see Section 3.2.6).

3. Multifunctional cloning vectors of S. cerevisiae

3.1 Customizing elements

3.1.1 *LacZα* fragment

Recombinant plasmid clones can be more easily identified after transformation of ligation mixtures into bacteria if the cloning vector expresses the *lacZα* fragment as a fusion protein encompassing the vector's cloning sites. Cloned DNA disrupts the *lacZα* coding sequence and recombinants can be distinguished from non-recombinants by a colony colour assay with X-gal in an *E. coli* host carrying a chromosomal *lacZΔM15* mutation.

3.1.2 f1 origin of DNA replication

The introduction of a short DNA fragment of about 460 bp from the filamentous phage f1 origin of DNA replication will enable yeast cloning vectors to produce single-stranded DNA molecules.(2). Single-stranded shuttle vector DNA may be useful for oligonucleotide-directed mutagenesis or for high quality DNA sequencing.

3.1.3 RNA polymerase promoters

Short oligonucleotides which function as phage RNA polymerase promoters can be positioned adjacent to the cloning sites of a yeast vector for rapid production of cRNA which can be used for *in vitro* protein synthesis studies or hybridization experiments. Both T3 and T7 RNA polymerase promoters have been incorporated into bacterial cloning vector backbones (pBluescript, see Chapter 8.).

3.1.4 Counterselectable yeast genes

Several yeast genes can be made conditionally and dominantly lethal (counter-selection) by simply changing the growth medium composition. Placing a counterselectable yeast gene within a yeast plasmid allows selection *against* the growth of yeast strains harbouring that particular episome. The most conveniently used counterselectable genes in *S. cerevisiae* are URA3, LYS2, CYH2, CAN1 (see Section 2.1.5). The counterselecting agents used in conjunction with these genes are 5-fluoroorotic acid (5-FOA), α-aminoadipic acid (α-AA), cycloheximide, and canavanine, respectively (25).

3.1.5 Colour marker

It is sometimes useful to be able to visually screen for the presence of a plasmid by viewing the colour of a growing yeast colony. The most commonly used colony colour system is based on suppression of a red pigment-producing allele of *ade2* (*ade2-101*) by a mutant tRNA (*SUP11*, *SUP4*). In this system, *ade2-101* is located within the yeast genome and the dominant t-RNA suppressor is located within a yeast shuttle vector. Strains containing a tRNA-marked shuttle vector produce white colonies on the appropriate growth medium and strains which have lost the vector grow as red pigmented colonies. A collection of shuttle vectors carrying both *SUP11* and *SUP4* genes have been constructed (23). Another useful colony colour system has been developed by positioning a wild-type *ADE2* gene within a shuttle vector (26).

3.1.6 Yeast promoters

Regulatable and constitutive promoters of varying strengths have been cloned into yeast episomes to create expression vectors for transcribing foreign or endogenous genes in yeast (27, 28).

The promoter of the *PHO5* gene (a yeast acid phosphatase) is inducible in growth medium containing low concentrations of phosphate (29). The *PHO5* promoter can also be activated by a temperature shift in special yeast strains which produce a temperature-sensitive transcriptional activator of *PHO5*, *PHO4* (27).

The promoter of the *PGK* gene (3-phosphoglycerate kinase) is constitutive and so active that it can produce greater than 1% of the total yeast cellular RNA (30). The promoter of the *ADHI*gene (alcohol dehydrogenase I) is constitutive but repressed to a variable extent when yeast use non-fermenting carbon sources such as lactate or acetate (31).

Several galactose metabolizing enzymes are transcribed from promoters that can be induced up to 1000-fold in the presence of galactose (32, 33). The divergent *GAL1* and *GAL10* gene promoters are probably the most well characterized *GAL* promoters, and small DNA fragments containing these two promoter elements have been used successfully for constructing many galactose inducible expression vectors (34).

The *CUP1* promoter, which is induced by the addition of copper to the growth medium, provides another choice for a regulatable promoter element (35). In one study, it was found that 13% of the total yeast cellular protein was composed of a nematode protein expressed from a *CUP1* promoter upon copper induction (36). Several *CUP1* expression vectors have been constructed (36, 37).

A novel approach to inducible expression systems involves the use of hybrid promoters containing glucocorticoid responsive elements (GRE) which constitute binding sites for the mammalian steroid receptor (38). After the addition of hormone, heterologously expressed mammalian steroid receptor activates transcription 50–100-fold from GRE-containing promoter constructs in yeast. Since there does not appear to be an endogenous yeast

Table 1. The pRS series of *S. cerevisiae* shuttle vectors

Plasmid	Selection	Features
YIp vectors (43)		
pRS303	HIS3	pBluescript backbone
pRS304	TRP1	pBluescript backbone
pRS305	LEU2	pBluescript backbone
pRS306	URA3	pBluescript backbone
YCp vectors (43)		
pRS313	HIS3	CEN6/ARSH4
pRS314	TRP1	CEN6/ARSH4
pRS315	LEU2	CEN6/ARSH4
pRS316	URA3	CEN6/ARSH4
YEp vectors (44)		
pRS323	HIS3	2μ ORI
pRS324	TRP1	2μ ORI
pRS325	LEU2	2μ ORI
pRS326	URA3	2μ ORI
Plasmid shuffle vectors (25)		
pRS317	LYS2	α-aminoadipate counterselection
pRS318	LEU2	CYH2—cycloheximide counterselection
pRS319	LEU2	CAN1—canavanine counterselection

steroid receptor, the effect of the mammalian hormone seems highly specific. Novel expression vectors containing steroid-inducible promoters have been designed (28).

3.1.7 *ori T*

Plasmid vectors can be transferred directly from *E. coli* to *S. cerevisiae* (and *S. pombe*) by transkingdom conjugation (transconjugation) if the mobilizing vector contains a small DNA element termed *ori T* (39, 40). An *ori T* sequence is found naturally within the bacterial plasmid pBR322 (nucleotides 2212–2353; reference 41), however, this sequence is disrupted in some pBR322 derivatives such as pUC and pBluescript. Of the few pBR322-based yeast shuttle vectors which have been tested for their ability to transconjugate (YEp13, YEp24, YCp50, and YIp5), only YEp13 does so with any reasonable efficiency (42). In order to make transconjugation a generally applicable procedure in *S. cerevisiae* we have recently constructed a new conjugal YCp vector formed by fusing pBR322 and the shuttle vector pRS313 (Sikorski *et al.*, unpublished data). This new vector, pRS196, can be transferred by simply mixing plasmid-containing bacteria with recipient yeast and selecting for HIS3$^+$ prototrophs (see *Protocol 4*). We have shown that pRS196 assumes in an intact circular form after mobilization (Sikorski *et al.*, unpublished data). Transkingdom conjugation may be particularly useful for transferring an array of episomes to a collection of yeast strains.

3.2 Selected cloning vectors of *S. cerevisiae*

3.2.1 pRS vector series

In an attempt to incorporate several customizing elements into a set of structurally similar yeast shuttle vectors we have created the pRS vector series (43, 44). To date, YCp, YIp, and YEp versions of pRS vectors have been created (*Table 1*). All members of the pRS series are based on the versatile backbone of pBluescript (2, see Chapter 8), so in addition to yeast-specific features they contain *lacZα*, f1 origin, and T3/T7 promoters. The extensive multiple cloning site of pRS vectors, which contains recognition sites for 18 different restriction enzymes, provides enough versatility for most cloning needs. Four different selectable yeast genes (*HIS3*, *TRP1*, *LEU2*, and *URA3*), are included in the pRS series of vectors.

3.2.2 Plasmid shuffle vectors

In some genetic manipulations there is a need to rapidly exchange or remove an entire gene from a yeast strain in one simple step. Such an exchange, or 'shuffle', can be easily accomplished by placing the gene of interest within an episomal cloning vector and exploiting the fact that episomes are naturally lost at a low frequency through mitotic missegregation. By incorporating a counterselectable gene into the episome, it is possible to directly select for growth of the desired, vector-less yeast strain (see Section 3.1.4). Yeast shuttle vectors exist which are compatible with plasmid shuffling (45). Recently we have constructed pRS-based shuffling vectors to facilitate cloning manipulations and offer a variety of counterselection methods (25).

3.2.3 λ-YES cloning

A unique λ phage-based cloning system has been developed for the efficient cloning of cDNA destined for expression in *S. cerevisiae* (46). A yeast shuttle vector carrying *URA3*, *CEN4*, *ARS1*, and the *GAL1* promoter was linearized and embedded into the non-essential region of phage λ. Recognition sites (*lox*) for the *cre* recombinase were included in the phage, flanking the inserted vector, and *cre* was engineered for constitutive expression in the *E. coli* host. Upon infection of a recombinant λ-YES phage library into the *cre*-expressing *E. coli*, a YCp vector is automatically excised from each phage and circularized into a plasmid form. Using this system a large amount of recombinant shuttle vector DNA can be readily isolated from transfected bacteria for subsequent transformation into yeast. cDNA libraries made in λ-YES should provide a valuable resource for the cloning of heterologous genes by direct complementation of yeast mutants.

3.2.4 Low copy number cloning—pMR366

MR366 was made from the *E. coli* plasmid pSC101, which has a stringent replication control, to allow the cloning of yeast genes which may be lethal in high copy number in *E. coli* (47). Perhaps as many as 25% of all yeast genes cloned into bacteria are expressed at some level from their natural yeast promoters. If the expressed yeast gene is lethal to *E. coli* then bacterial clones containing the gene of interest can never be obtained. Compounding the potential problem is the fact that virtually all yeast–*E. coli* shuttle vectors attain a relatively high copy number in bacteria. The regulated mode of DNA replication of MR366 ensures that a low plasmid copy number is maintained in bacteria so that the deleterious effects of any toxic yeast inserts should be minimized.

3.2.5 Transcription-based detection systems

Advantage has been taken of the segmental nature of the *GAL4* transcription activator protein to construct vector systems for the detection of protein–protein (48) and protein–DNA interactions (49). It has been shown that two coexpressed fusion proteins, one containing the *GAL4* activator and the other the *GAL4* DNA-binding domain, can activate transcription from a target sequence if they form a noncovalent association (50). By constructing DNA-binding and activator fusions to polypeptides that physically interact, activation of a reporter gene (i.e. *lacZ*) can be used as an assay of *in vivo* protein–protein interaction. One can also fuse open reading frames to the *GAL4* activator and screen or select in yeast for functional DNA-binding domains using a reporter gene whose promoter contains any chosen target sequence (51).

3.2.6 YAC cloning

Large YLp constructs have been made by *in vitro* ligation followed by transformation into yeast to produce stable yeast artificial chromosomes (YACs) containing hundreds of kilobases of passenger DNA (24). To produce a YAC capable of propagation in yeast the following elements must be ligated to the ends of the passenger DNA: two telomeres, one *ARS*, and one or two selectable yeast genes. In practice, the required DNA elements are derived from the sub-fragments of a single plasmid (52) or from two entirely separate plasmids (53). Although the current methodology of YAC cloning is quite laborious (54), several groups have succeeded in constructing large genomic libraries from which entire transcription units or genetic loci can be cloned and analysed (55, 56).

4. Yeast host strain

The minimal requirement for a yeast strain to serve as a host for a linear or circular plasmid vector is a genetic background compatible with the mode of vector selection. For biosynthetic marker genes, the host should preferably carry a non-revertable mutant allele of the gene under selection. Mutations which inactivate the chromosomal *URA3* or *LYS2* loci can be produced *de novo* by counterselection (see Section 3.1.4). A more general approach is to introduce non-reverting auxotrophic markers into the yeast genome through targeted transformation using specially designed yeast integrating plasmids (YIp). Vectors have been designed to target deletion mutations to the *HIS3*, *TRP1*, *URA3*, and *LEU2* loci (43). Genetic crosses can be performed to quickly introduce multiple auxotrophic mutations into a host strain background.

5. Introduction of DNA into S. cerevisiae

Techniques of various degrees of efficiency can be used to introduce DNA into *S. cerevisiae*. Exposure to lithium salts makes yeast competent to transform episomal DNA with a yield of about 10^3 colonies per microgram input DNA. An electroporation device can be used to increase the efficiency to 1–5 $\times 10^5$ colonies per microgram input DNA. Alternatively, degradative enzymes can be used to remove the yeast cell wall and produce spheroplasts competent to transform episomes at 1×10^4–10^5 colonies per microgram input DNA. A novel method for introducing plasmid DNA into yeast is the direct transfer of episomal vectors from donor bacteria to yeast by transkingdom conjugation, a process which can produce about 100 transconjugants per patch of yeast on solid medium.

Protocol 1. Lithium acetate-mediated transformation[a]

Materials

- YPD broth: 1% bacto yeast extract, 2% bactopeptone, 2% dextrose
- 0.1 M lithium acetate (LiOAc)
- herring or salmon sperm DNA
- 40% PEG 4000 in 10 mM Tris pH 7.5, 0.1 M LiOAc

- distilled water (dH$_2$O)
- SD medium: 0.67% bacto yeast nitrogen base (without amino acids), 2% dextrose, 0.5% ammonium sulfate, 2% bacto agar, and supplements as required by the chosen yeast strain[b]

Method

1. Dilute a saturated yeast culture 1:100 in 50 ml of YPD. This will yield enough competent cells for ten transformations.

2. Incubate with vigorous shaking at 30°C until an OD$_{600}$ of 1–2 is reached (approximately 5 h).

3. Collect cells by centrifugation in a table-top centrifuge (Sorvall, 3500 r.p.m. for 5 min).

4. Wash cells once with 20 ml of 0.1 M LiOAc and collect by centrifugation as in step **3**.

5. Resuspend cells in 10 ml 0.1 M LiOAc and incubate for 1 h at 30°C with gentle mixing. Cells can be stored for up to two weeks at this point at 4°C prior to use.

6. Collect cells by centrifugation as in step **3**, resuspend in 0.5 ml 0.1 M LiOAc, and dispense 50 μl aliquots into 1.5 ml microcentrifuge tubes.

7. Add 1–10 μg of yeast episomal vector DNA, and 5 μg of heat-denatured herring or salmon sperm carrier DNA to an aliquot of cells, and incubate at 30°C for 10 min. Note that the carrier DNA should be pre-treated by shearing through a 25 gauge needle to a 5–15 kb size range before use.

8. Add 0.5 ml of 40% PEG 4000 (in 10 mM Tris pH 7.5, 0.1 M LiOAc). The pH of this solution should be checked periodically and the solution replaced if acidic.

9. Incubate at 30°C for 60 min.

10. Heat shock at 42°C for 5 min.

11. Add 1.0 ml dH$_2$O, mix, and pellet cells for 5 sec in a microfuge.

12. Wash once with 1.0 ml dH$_2$O and repellet cells as in step **11**.

13. Resuspend the cells in 0.1 ml dH$_2$O and spread them on to solid SD

medium appropriate for the selectable yeast gene within the cloning vector.

[a] Adapted from Ito, H., Fukuda, Y., Murata, K., and Kimura, A. (1983). *J. Bacteriol.*, **153**, 163.
[b] Sherman, F. (1991). In *Methods in Enzymology* (ed. C Guthrie and G.R. Fink), Vol. 194, pp. 3–21.

Protocol 2. Electroporation [a]

Materials

- YPD broth (see *Protocol 1*)
- 0.1 M sorbitol
- TE buffer: 10 mM Tris pH 8.0, 1 mM EDTA
- Gene Pulsar apparatus (Bio-Rad)
- 0.2 cm electroporator cuvettes (Bio-Rad)
- SD medium (see *Protocol 1*) containing 1.0 M sorbitol
- distilled water (dH$_2$O)

Method

1. Dilute a saturated yeast culture grown in YPD 1:100 into 500 ml of fresh YPD. Incubate at 30°C with agitation to an OD$_{600}$ of 1.3–1.5.

2. Collect cells by centrifugation for 5 min at 5000 r.p.m. in a Sorvall GSA rotor. From this step on, all solutions and centrifugations should be at 4°C.

3. Resuspend cells in 500 ml dH$_2$O and pellet as in step **2**.

4. Resuspend cells in 250 ml dH$_2$O and pellet as in step **2**.

5. Resuspend cells in 20 ml of 1 M sorbitol and collect by centrifugation at 5000 r.p.m. for 5 min in a Sorvall SS34 rotor.

6. Resuspend cells in 0.5 ml of 1 M sorbitol.

7. Add 40 μl of prepared yeast cells from step **6** to about 100 ng of vector DNA (in less than 5 μl of TE buffer) and mix in a microcentrifuge tube. Keep on ice.

8. Transfer the yeast–DNA mix to a 0.2 cm electroporation cuvette and insert the cuvette into the electroporator. We use the Gene Pulser apparatus by Bio-Rad.

9. Apply one pulse at the following settings: 1.5 kV, 25 μF, and 200 Ω.

10. Add 1.0 ml cold 1 M sorbitol to the cuvette and mix.

11. Plate the electroporation mix on to SD medium which contains 1.0 M sorbitol and incubate at 30°C. Note that top agar is not required.

[a] Adapted from Becker, D. M. and Guarente, L. (1991). In *Methods in Enzymology* (ed. C. Guthrie and G. R. Fink), Vol. 194, pp. 182–87.

Protocol 3. Spheroplast transformation[a]

Materials

- YPD broth (see *Protocol 1*)
- distilled water (dH$_2$O)
- SPEM: 1 M sorbitol, 10 mM Na phosphate pH 7.5, 10 mM EDTA, 30 mM β-mercaptoethanol (add β-mercaptoethanol just before use)
- 1 M sorbitol
- 10 mg/ml 20T zymolyase in 10 mM Na phosphate, pH 7.5 (ICN Immunochemicals)
- 5% SDS

- STC: 1 M sorbitol, 10 mM Tris pH 7.5, 10 mM CaCl$_2$
- PEG solution: 20% PEG 8000, 10 mM Tris pH 7.5, 10 mM CaCl$_2$
- SOS: 1 M sorbitol, 6.5 mM CaCl$_2$, 0.25% yeast extract, and 0.5% bactopeptone
- TOP agar: molten SD medium (see *Protocol 1*) with 1 M sorbitol at 48°C
- SORB medium: solid SD medium (see *Protocol 1*) with 1 M sorbitol

Method

1. Add 5 ml of a fresh overnight culture of yeast to 50 ml of YPD broth and culture with agitation at 30°C to an OD$_{600}$ of 4.0.

2. Collect cells by centrifugation in a table-top Sorvall centrifuge for 5 min at 3000 r.p.m.

3. Wash cells once with 20 ml dH$_2$O and pellet as in step **2**.

4. Wash cells once with 20 ml of 1 M sorbitol and pellet as in step **2**.

5. Resuspend cells in 20 ml SPEM.

6. Add 45 μl of 10 mg/ml 20T zymolyase and incubate with gently shaking at 30°C for approximately 20 min. The extent of cell wall digestion is critical and can be gauged by adding a small sample of cells to an equal volume of 5 % SDS and observing the formation of spheroplast 'ghosts' by phase-contrast microscopy. About 90% of the cells should be ghosts for good transformation. It may be necessary to titrate the amount of zymolyase required for best results.

7. Collect the cells by centrifugation as in step **2**.

8. Gently resuspend the spheroplasts in 20 ml of 1 M sorbitol and pellet as in step **2**. Wash cells once with 20 ml STC and pellet as in step **2**.

9. Resuspend spheroplasts in 2 ml STC. They are stable in this form for at least 1 h at room temperature.

10. To 100 μl of spheroplasts add 1 ng of plasmid DNA (in a volume of less than 10 μl). Incubate for 10 min at room temperature.

11. Add 1 ml PEG solution, mix gently, and incubate for 10 min at room temperature. Pellet as in step **2**.

12. Resuspend cells in 150 μl SOS and incubate at 30°C for 30–40 min without shaking.

13. Gently resuspend the spheroplasts, add them to 6–8 ml TOP agar (at 48°C), and pour on to selective SORB agar plates. Incubate at 30°C to form colonies.

[a] Adapted from McCormick, M.K., Shero, J.H., Connelly, C., Antonarakis, S.A., and Hieter, P. (1990). *Technique*, **2**, 65.

Protocol 4. Transkingdom conjugation on solid medium[a]

Materials

- bacterial strain JB 117[b] [*SupE44, thi-1, thr-1, leuB6, lacy1, tonA21,* λ−, (*rk+, mk+*), (*mcra−, mcrb+*), (pDPT51-Tmp[r]-Ap[r])]
- LB broth: 1% bactotryptone, 0.5% bacto yeast extract, 1% NaCl
- trimethoprim and tetracycline

- TNB: 50 mM Tris pH 7.6, 0.05% NaCl
- YPD solid medium: YPD (see *Protocol 1*) plus 2% agar
- SD medium (see *Protocol 1*)

Method

1. Grow a culture of JB117 (harbouring the plasmid to be mobilized) to saturation in LB broth plus trimethoprim (200 μg/ml) and the appropriate selection for the mobilizing plasmid. If the mobilizing plasmid is pRS196 (Tc[r]) use tetracycline at 12.5 μg/ml. Note that a 'helper' plasmid (pDPT51) in JB117, which provides *mob* and *tra* functions essential for transconjugation, carries trimethoprim and ampicillin resistance genes for selection.
2. Collect the bacteria at 4000 *g* for 10 min, and resuspend in 100 ml TNB.
3. Collect the bacteria as in step **2** and resuspend in 1 ml TNB. The cells, now concentrated 100-fold, can be used immediately or frozen for further use after the addition of 7.5% DMSO.
4. Patch the recipient yeast strain to YPD in grids of about 1 sq. cm one day prior to use. Incubate overnight at 30°C.
5. Pipette 10 μl of donor bacteria from step **3** directly on to each yeast patch. Incubate overnight at 30°C to allow plasmid transfer.
6. Using a velvet cloth replica device, print the patches on to SD minimal media and incubate at 30°C. When using pRS196 (*HIS3*), select transconjugants on medium lacking histidine. Note that the donor bacteria JB117 is a leucine and threonine auxotroph and bacterial overgrowth can be prevented by excluding leucine or threonine from the yeast growth medium.

[a] Sikorski, R. S., Michaud, W., and Hieter, P. (unpublished data).
[b] Heinemann, J. A. and Sprague, G. F. (1989). *Nature*, **340**, 205.

Acknowledgements

I would like to thank Phil Hieter for his constant support and encouragement during the construction of pRS vectors and Jef Boeke for his endless enthusiasm during our many collaborations.

References

1. Yanisch-Perron, C., Vierira, J., and Messing, J. (1985). *Gene*, **33**, 103.
2. Short, J. M., Fernandez, J. M., Sorge, J. A., and Huse, W. D. (1988). *Nucleic. Acids Res.*, **15**, 7583
3. Houten, J. V. and Newlon, C. S. (1990). *Mol. Cell. Biol.*, **10**, 3917.
4. Deshpande, A. M. and Newlon, C. S. (1992). *Mol. Cell. Biol.*, **12**, 4305.
5. Marahrens, Y. and Stillman, B. (1992). *Science*, **255**, 817.
6. Hieter, P., Pridmore, D., Hegemann, J., Thomas, M., Davis, R., and Philippsen, P. (1985). *Cell*, **42**, 913.
7. Cottarel, G., Shero, J. H., Hieter, P., and Hegeman, J. H. (1989). *Mol. Cell. Biol.*, **9**, 3342.
8. Dobson, M. J., Yull, F. E., Kingsman, S. M., and Kingsman, A. J. (1988). *Nucleic Acids Res.*, **16**, 7103.
9. Som, T., Armstrong, K. A., Volkert, F. C., and Broach, J. (1988). *Cell*, **52**, 27.
10. Struhl, K. (1985). *Nucleic Acids Res.*, **13**, 8587.
11. Tschumper, G. and Carbon, J. (1980). *Gene*, **10**, 157.
12. Zakian, V. A. and Scott, J. F. (1982). *Mol. Cell. Biol.*, **2**, 221.
13. Andreadis, A., Hsu, Y. P., Kohlhaw, G. B., and Schimmel, P. (1982). *Cell*, **31**, 319.
14. Rose, A. B. and Broach, J. R. (1990). In *Methods in enzymology* (ed. D.V. Goedel), Vol. 185, pp. 234–79.
15. Rose, M., Grisafi, P., and Botstein, D. (1984). *Gene*, **29**, 113.
16. Boeke, J. D., Lacroute, F., and Fink, G. (1984). *Mol. Gen. Genet.*, **197**, 345.
17. Barnes, D. A. and Thorner, J. (1986). *Mol. Cell. Biol.*, **6**, 2828.
18. Chatoo, B. B., Sherman, F., Azubalis, D. A., Fjellstedt, T. A., Mehnert, D., and Ogur, M. (1979). *Genetics*, **93**, 51.
19. Wang, S. S. and Zakian, V. A. (1990). *Mol. Cell. Biol.*, **10**, 4415.
20. Rose, M. D. and Broach, J. R. (1991). In *Methods in enzymology* (ed. C. Guthrie and G. R. Fink), Vol. 194, pp. 195–230. Academic Press, London.
21. Bonneaud, N., Ozier-Kalogeupoulos, O., Li, G. Y., Labouesse, M., Minielle-Sebastia, L., and Lacroute, F. (1991). *Yeast*, **7**, 609.
22. Vernet, T., Dignard, D., and Thomas, D. Y. (1987). *Gene*, **52**, 225.
23. Elledge, S. J. and Davis, R. W. (1988). *Gene*, **70**, 303.
24. Murray, A. W. and Szostak, J. W. (1987). *Sci. Am.*, **257**, 62.
25. Sikorski, R. S. and Boeke, J. D. (1991). In *Methods in enzymology* (ed. C. Guthrie and G. R. Fink), Vol. 194, pp. 302–18. Academic Press, London.
26. Stotz, A. and Linder, P. (1990). *Gene*, **95**, 91.
27. Schneider, J. C. and Guarente, L. (1991). In *Methods in enzymology* (ed. C. Guthrie and G. R. Fink), Vol. **194**, pp. 373–88. Academic Press, London.

28. Schena, M., Picard, D., and Yamamoto, K. (1991). In *Methods in enzymology* (ed. C. Guthrie and G. R. Fink), Vol. 194, pp. 389–98. Academic Press, London.
29. Rogers, D. T., Levine, J. M., and Bostian, K. A. (1982). *Proc. Natl. Acad. Sci. USA*, **79**, 2157.
30. Holland, M. J. and Holland, J. P. (1978). *Biochemistry*, **17**, 4900.
31. Denis, C. L., Ferguson, J., and Young, E. T. (1983). *J. Biol. Chem.*, **258**, 1165.
32. Guarente, L., Yocum, R. R., and Gifford, P. (1982). *Proc. Natl. Acad. Sci. USA*, **79**, 7410.
33. Johnston, M. (1987). *Microbiol. Rev.*, **51**, 458.
34. St. John, T. P. and Davis, R. W. (1981). *J. Mol. Biol.*, **152**, 285.
35. Macreadie, I. G., Jagadish, M. N., Azad, A. A., and Vaugham, P. R. (1989). *Plasmid*, **21**, 147.
36. Macreadie, I. G., Horaitis, O., Herkuylen, A. J., and Savin, K. W. (1991). *Gene*, **104**, 107.
37. Macreadie, I. G. (1990). *Nucleic Acids Res.*, **18**, 1078.
38. Metzger, D., White, J. H., and Chambon, P. (1988).*Nature*, **334**, 31.
39. Heinemann, J. A. and Sprague, G. F. (1989). *Nature*, **340**, 205.
40. Sikorski, R. S., Michaud, W., Levine, H., Boeke, J. D., and Hieter, P. (1990). *Nature*, **345**, 581.
41. Finnegan, J. and Sherratt, D. (1982). *Mol. Gen. Genet.*, **185**, 344.
42. Heinemann, J. A. and Sprague, G.F. (1991). In *Methods in enzymology* (ed. C. Guthrie and G. R. Fink), Vol. 194, pp. 187–95. Academic Press, London.
43. Sikorski, R. S. and Hieter, P. (1989). *Genetics*, **122**, 19.
44. Christianson, T. W., Sikorski, R. S., Dante, M., Shero, J. H., and Hieter, P. (1992). *Gene*, **110**, 119.
45. Boeke, J. B., Trueheart, J., Natsoulis, G., and Fink, G. R. (1988). In *Methods in enzymology* (ed. R. Wu and L. Grossman), Vol. 154, pp. 164–75. Academic Press, London.
46. Elledge, S. J., Mulligan, J. T., Ramer, S. W., Spottswood, M., and Davis, R. W. (1991). *Proc. Natl. Acad. Sci. USA*, **88**, 1731.
47. Rose, M. D., Misra, L. M., and Vogel, J. P. (1989). *Cell*, **57**, 1211.
48. Chevray, P. M. and Nathans, D. (1992). *Proc. Natl. Acad. Sci. USA*, **89**, 5789.
49. Bonner, J. J. (1991). *Gene*, **104**, 113.
50. Fields, S. and Song, O. K. (1989).*Nature*, **340**, 245.
51. Fields, S. and Jang, S. K. (1990). *Science*, **249**, 1046.
52. Burke, D. T., Carle, G. F., and Olson, C. M. (1987). *Science*, **236**, 806.
53. Shero, J. H., McCormick, M. K., Antonarakis, S. E., and Hieter, P. (1991). *Genomics*, **10**, 505.
54. Burke, D. T. and Olson, M. V. (1991). In *Methods in enzymology* (ed. C. Guthrie and G. R. Fink), Vol. 194, pp. 251–70. Academic Press, London.
55. Albertsen, H. M., Abderrahim, H., Cann, H. M., Dausset, J., Le Paslier, D., and Cohen, D. (1990). *Proc. Natl. Acad. Sci. USA* **87**, 4256.
56. Larin, Z., Monaco, A. P., and Lehrach, H. (1991). *Proc. Natl. Acad. Sci. USA*, **88**, 4123.

Addresses of suppliers

Amersham Corporation, 2636S Clearbrook Drive, Arlington Heights, Illinois 60005, USA.

Amersham International plc, Amersham Place, Little Chalfont, Buckinghamshire, HP7 9NA, UK.

Bachem AG, Hauptstrasse 144, 4416 Bubendorf, Switzerland.

Beckman Instruments Inc., Spinco Division, 1050 Page Mill Road, Palo Alto, CA 94304, USA.

Beckman Ltd, Turnpike Road, Cressex Industrial Estate, High Wycombe, HP12 3NR, UK.

Becton Dickinson Immunocytometry Systems, PO Box 7375, Mountain View, CA 94039, USA.

Bethesda Research Laboratories, Life Technologies Inc., PO Box 6009, Gaithersburg, MD 20877, USA; Life Technologies Ltd, PO Box 35, Trident House, Renfrew Road, Paisley, PA3 4EF, UK.

BIO 101 Inc., PO Box 2284, La Jolla, CA 92038, USA.

Bio-Rad, 2200 Wright Avenue, Richmond, CA 94804, USA.

BRL—see **Bethesda Research Laboratories**.

Difco Laboratories, PO Box 331058, Detroit, MI 48232, USA; PO Box 14B, Central Avenue, East Molesey, Surrey, KT8 08E, UK.

DuPont Company, Biotechnology Systems, BRML, G-50636, Wilmington, DE 19898, USA.

Falcon—see **Becton Dickinson**.

ICN Biomedicals Inc., ICN Plaza, 3300 Hyland Avenue, Costa Mesa, CA 92626, USA.

ICN Biomedicals Ltd, Eagle House, Peregrine Business Park, Gomm Road, High Wycombe, Buckinghamshire, HP13 7DL, UK.

Millipore Corporation, PO Box 255, Bedford, MA 01730, USA.

Millipore UK Ltd, 11–15 Peterborough Road, Harrow, Middlesex, HA1 2YH, UK.

New England Biolabs Inc., 32 Tozer Road, Beverly, MA 01915, USA.

Pharmacia LKB Biotechnology AB, Björkgatan 30, 751 82 Uppsala, Sweden.

Pharmacia LKB Biotechnology Inc., 800 Centennial Avenue, PO Box 1327, Piscataway, NJ 08855, USA.

Promega Biotech, 2800 S. Fish Hatchery Road, Adison, WI 53711, USA.

Quiagen Inc., 9259 Eton Avenue, Chatsworth, CA 91311, USA.

Carl Roth, Schloempertenstrasse 1–5, 7500 Karlsrue, 2121, Germany.
Serva Feinbiochemica GmbH, Postfach 105260, 6900 Heidelberg, Germany.
Sigma Chemical Co., PO Box 14508, St Louis, MO 63178, USA.
Sorvall—see **DuPont.**
Stratagene, 11099 North Torrey Pines Road, La Jolla, CA 92037, USA.

Index

ORDER OTHER TITLES OF INTEREST TODAY

Price list for: UK, Europe, Rest of World (excluding US and Canada)

138. **Plasmids (2/e)** Hardy, K.G. (Ed)
...... Spiralbound hardback 0-19-963445-9 **£30.00**
...... Paperback 0-19-963444-0 **£19.50**
136. **RNA Processing: Vol. II** Higgins, S.J. & Hames, B.D. (Eds)
...... Spiralbound hardback 0-19-963471-8 **£30.00**
...... Paperback 0-19-963470-X **£19.50**
135. **RNA Processing: Vol. I** Higgins, S.J. & Hames, B.D. (Eds)
...... Spiralbound hardback 0-19-963344-4 **£30.00**
...... Paperback 0-19-963343-6 **£19.50**
134. **NMR of Macromolecules** Roberts, G.C.K. (Ed)
...... Spiralbound hardback 0-19-963225-1 **£32.50**
...... Paperback 0-19-963224-3 **£22.50**
133. **Gas Chromatography** Baugh, P. (Ed)
...... Spiralbound hardback 0-19-963272-3 **£40.00**
...... Paperback 0-19-963271-5 **£27.50**
132. **Essential Developmental Biology** Stern, C.D. & Holland, P.W.H. (Eds)
...... Spiralbound hardback 0-19-963423-8 **£30.00**
...... Paperback 0-19-963422-X **£19.50**
131. **Cellular Interactions in Development** Hartley, D.A. (Ed)
...... Spiralbound hardback 0-19-963391-6 **£30.00**
...... Paperback 0-19-963390-8 **£18.50**
129 **Behavioural Neuroscience: Volume II** Sahgal, A. (Ed)
...... Spiralbound hardback 0-19-963458-0 **£32.50**
...... Paperback 0-19-963457-2 **£22.50**
128 **Behavioural Neuroscience: Volume I** Sahgal, A. (Ed)
...... Spiralbound hardback 0-19-963368-1 **£32.50**
...... Paperback 0-19-963367-3 **£22.50**
127. **Molecular Virology** Davison, A.J. & Elliott, R.M. (Eds)
...... Spiralbound hardback 0-19-963358-4 **£35.00**
...... Paperback 0-19-963357-6 **£25.00**
126. **Gene Targeting** Joyner, A.L. (Ed)
...... Spiralbound hardback 0-19-963407-6 **£30.00**
...... Paperback 0-19-9634036-8 **19.50**
125. **Glycobiology** Fukuda, M. & Kobata, A. (Eds)
...... Spiralbound hardback 0-19-963372-X **£32.50**
...... Paperback 0-19-963371-1 **£22.50**
124. **Human Genetic Disease Analysis (2/e)** Davies, K.E. (Ed)
...... Spiralbound hardback 0-19-963309-6 **£30.00**
...... Paperback 0-19-963308-8 **£18.50**
122. **Immunocytochemistry** Beesley, J. (Ed)
...... Spiralbound hardback 0-19-963270-7 **£35.00**
...... Paperback 0-19-963269-3 **£22.50**
123. **Protein Phosphorylation** Hardie, D.G. (Ed)
...... Spiralbound hardback 0-19-963306-1 **£32.50**
...... Paperback 0-19-963305-3 **£22.50**
121. **Tumour Immunobiology** Gallagher, G., Rees, R.C. & others (Eds)
...... Spiralbound hardback 0-19-963370-3 **£40.00**
...... Paperback 0-19-963369-X **£27.50**
120. **Transcription Factors** Latchman, D.S. (Ed)
...... Spiralbound hardback 0-19-963342-8 **£30.00**
...... Paperback 0-19-963341-X **£19.50**
119. **Growth Factors** McKay, I. & Leigh, I. (Eds)
...... Spiralbound hardback 0-19-963360-6 **£30.00**
...... Paperback 0-19-963359-2 **£19.50**
118. **Histocompatibility Testing** Dyer, P. & Middleton, D. (Eds)
...... Spiralbound hardback 0-19-963364-9 **£32.50**
...... Paperback 0-19-963363-0 **£22.50**

117. **Gene Transcription** Hames, B.D. & Higgins, S.J. (Eds)
...... Spiralbound hardback 0-19-963292-8 **£35.00**
...... Paperback 0-19-963291-X **£25.00**
116. **Electrophysiology** Wallis, D.I. (Ed)
...... Spiralbound hardback 0-19-963348-7 **£32.50**
...... Paperback 0-19-963347-9 **£22.50**
115. **Biological Data Analysis** Fry, J.C. (Ed)
...... Spiralbound hardback 0-19-963340-1 **£50.00**
...... Paperback 0-19-963339-8 **£27.50**
114. **Experimental Neuroanatomy** Bolam, J.P. (Ed)
...... Spiralbound hardback 0-19-963326-6 **£32.50**
...... Paperback 0-19-963325-8 **£22.50**
113. **Preparative Centrifugation** Rickwood, D. (Ed)
...... Spiralbound hardback 0-19-963208-1 **£45.00**
...... Paperback 0-19-963211-1 **£25.00**
...... Paperback 0-19-963099-2 **£25.00**
112. **Lipid Analysis** Hamilton, R.J. & Hamilton, Shiela (Eds)
...... Spiralbound hardback 0-19-963098-4 **£35.00**
...... Paperback 0-19-963099-2 **£25.00**
111. **Haemopoiesis** Testa, N.G. & Molineux, G. (Eds)
...... Spiralbound hardback 0-19-963366-5 **£32.50**
...... Paperback 0-19-963365-7 **£22.50**
110. **Pollination Ecology** Dafni, A.
...... Spiralbound hardback 0-19-963299-5 **£32.50**
...... Paperback 0-19-963298-7 **£22.50**
109. **In Situ Hybridization** Wilkinson, D.G. (Ed)
...... Spiralbound hardback 0-19-963328-2 **£30.00**
...... Paperback 0-19-963327-4 **£18.50**
108. **Protein Engineering** Rees, A.R., Sternberg, M.J.E. & others (Eds)
...... Spiralbound hardback 0-19-963139-5 **£35.00**
...... Paperback 0-19-963138-7 **£25.00**
107. **Cell-Cell Interactions** Stevenson, B.R., Gallin, W.J. & others (Eds)
...... Spiralbound hardback 0-19-963319-3 **£32.50**
...... Paperback 0-19-963318-5 **£22.50**
106. **Diagnostic Molecular Pathology: Volume I** Herrington, C.S. & McGee, J. O'D. (Eds)
...... Spiralbound hardback 0-19-963237-5 **£30.00**
...... Paperback 0-19-963236-7 **£19.50**
105. **Biomechanics-Materials** Vincent, J.F.V. (Ed)
...... Spiralbound hardback 0-19-963223-5 **£35.00**
...... Paperback 0-19-963222-7 **£25.00**
104. **Animal Cell Culture (2/e)** Freshney, R.I. (Ed)
...... Spiralbound hardback 0-19-963212-X **£30.00**
...... Paperback 0-19-963213-8 **£19.50**
103. **Molecular Plant Pathology: Volume II** Gurr, S.J., McPherson, M.J. & others (Eds)
...... Spiralbound hardback 0-19-963352-5 **£32.50**
...... Paperback 0-19-963351-7 **£22.50**
102 **Signal Transduction** Milligan, G. (Ed)
...... Spiralbound hardback 0-19-963296-0 **£30.00**
...... Paperback 0-19-963295-2 **£18.50**
101. **Protein Targeting** Magee, A.I. & Wileman, T. (Eds)
...... Spiralbound hardback 0-19-963206-5 **£30.00**
...... Paperback 0-19-963210-3 **£22.50**
100. **Diagnostic Molecular Pathology: Volume II: Cell and Tissue Genotyping** Herrington, C.S. & McGee, J.O'D. (Eds)
...... Spiralbound hardback 0-19-963239-1 **£30.00**
...... Paperback 0-19-963238-3 **£19.50**
99. **Neuronal Cell Lines** Wood, J.N. (Ed)
...... Spiralbound hardback 0-19-963346-0 **£32.50**
...... Paperback 0-19-963345-2 **£22.50**

98. **Neural Transplantation** Dunnett, S.B. & Björklund, A. (Eds)
...... Spiralbound hardback 0-19-963286-3 **£30.00**
...... Paperback 0-19-963285-5 **£19.50**
97. **Human Cytogenetics: Volume II: Malignancy and Acquired Abnormalities (2/e)** Rooney, D.E. & Czepulkowski, B.H. (Eds)
...... Spiralbound hardback 0-19-963290-1 **£30.00**
...... Paperback 0-19-963289-8 **£22.50**
96. **Human Cytogenetics: Volume I: Constitutional Analysis (2/e)** Rooney, D.E. & Czepulkowski, B.H. (Eds)
...... Spiralbound hardback 0-19-963288-X **£30.00**
...... Paperback 0-19-963287-1 **£22.50**
95. **Lipid Modification of Proteins** Hooper, N.M. & Turner, A.J. (Eds)
...... Spiralbound hardback 0-19-963274-X **£32.50**
...... Paperback 0-19-963273-1 **£22.50**
94. **Biomechanics-Structures and Systems** Biewener, A.A. (Ed)
...... Spiralbound hardback 0-19-963268-5 **£42.50**
...... Paperback 0-19-963267-7 **£25.00**
93. **Lipoprotein Analysis** Converse, C.A. & Skinner, E.R. (Eds)
...... Spiralbound hardback 0-19-963192-1 **£30.00**
...... Paperback 0-19-963231-6 **£19.50**
92. **Receptor-Ligand Interactions** Hulme, E.C. (Ed)
...... Spiralbound hardback 0-19-963090-9 **£35.00**
...... Paperback 0-19-963091-7 **£27.50**
91. **Molecular Genetic Analysis of Populations** Hoelzel, A.R. (Ed)
...... Spiralbound hardback 0-19-963278-2 **£32.50**
...... Paperback 0-19-963277-4 **£22.50**
90. **Enzyme Assays** Eisenthal, R. & Danson, M.J. (Eds)
...... Spiralbound hardback 0-19-963142-5 **£35.00**
...... Paperback 0-19-963143-3 **£25.00**
89. **Microcomputers in Biochemistry** Bryce, C.F.A. (Ed)
...... Spiralbound hardback 0-19-963253-7 **£30.00**
...... Paperback 0-19-963252-9 **£19.50**
88. **The Cytoskeleton** Carraway, K.L. & Carraway, C.A.C. (Eds)
...... Spiralbound hardback 0-19-963257-X **£30.00**
...... Paperback 0-19-963256-1 **£19.50**
87. **Monitoring Neuronal Activity** Stamford, J.A. (Ed)
...... Spiralbound hardback 0-19-963244-8 **£30.00**
...... Paperback 0-19-963243-X **£19.50**
86. **Crystallization of Nucleic Acids and Proteins** Ducruix, A. & Giegé, R. (Eds)
...... Spiralbound hardback 0-19-963245-6 **£35.00**
...... Paperback 0-19-963246-4 **£25.00**
85. **Molecular Plant Pathology: Volume I** Gurr, S.J., McPherson, M.J. & others (Eds)
...... Spiralbound hardback 0-19-963103-4 **£30.00**
...... Paperback 0-19-963102-6 **£19.50**
84. **Anaerobic Microbiology** Levett, P.N. (Ed)
...... Spiralbound hardback 0-19-963204-9 **£32.50**
...... Paperback 0-19-963262-6 **£22.50**
83. **Oligonucleotides and Analogues** Eckstein, F. (Ed)
...... Spiralbound hardback 0-19-963280-4 **£32.50**
...... Paperback 0-19-963279-0 **£22.50**
82. **Electron Microscopy in Biology** Harris, R. (Ed)
...... Spiralbound hardback 0-19-963219-7 **£32.50**
...... Paperback 0-19-963215-4 **£22.50**
81. **Essential Molecular Biology: Volume II** Brown, T.A. (Ed)
...... Spiralbound hardback 0-19-963112-3 **£32.50**
...... Paperback 0-19-963113-1 **£22.50**
80. **Cellular Calcium** McCormack, J.G. & Cobbold, P.H. (Eds)
...... Spiralbound hardback 0-19-963131-X **£35.00**
...... Paperback 0-19-963130-1 **£25.00**
79. **Protein Architecture** Lesk, A.M.
...... Spiralbound hardback 0-19-963054-2 **£32.50**
...... Paperback 0-19-963055-0 **£22.50**
78. **Cellular Neurobiology** Chad, J. & Wheal, H. (Eds)
...... Spiralbound hardback 0-19-963106-9 **£32.50**
...... Paperback 0-19-963107-7 **£22.50**
77. **PCR** McPherson, M.J., Quirke, P. & others (Eds)
...... Spiralbound hardback 0-19-963226-X **£30.00**
...... Paperback 0-19-963196-4 **£19.50**
76. **Mammalian Cell Biotechnology** Butler, M. (Ed)
...... Spiralbound hardback 0-19-963207-3 **£30.00**
...... Paperback 0-19-963209-X **£19.50**
75. **Cytokines** Balkwill, F.R. (Ed)
...... Spiralbound hardback 0-19-963218-9 **£35.00**
...... Paperback 0-19-963214-6 **£25.00**

74. **Molecular Neurobiology** Chad, J. & Wheal, H. (Eds)
...... Spiralbound hardback 0-19-963108-5 **£30.00**
...... Paperback 0-19-963109-3 **£19.50**
73. **Directed Mutagenesis** McPherson, M.J. (Ed)
...... Spiralbound hardback 0-19-963141-7 **£30.00**
...... Paperback 0-19-963140-9 **£19.50**
72. **Essential Molecular Biology: Volume I** Brown, T.A. (Ed)
...... Spiralbound hardback 0-19-963110-7 **£32.50**
...... Paperback 0-19-963111-5 **£22.50**
71. **Peptide Hormone Action** Siddle, K. & Hutton, J.C.
...... Spiralbound hardback 0-19-963070-4 **£32.50**
...... Paperback 0-19-963071-2 **£22.50**
70. **Peptide Hormone Secretion** Hutton, J.C. & Siddle, K. (Eds)
...... Spiralbound hardback 0-19-963068-2 **£35.00**
...... Paperback 0-19-963069-0 **£25.00**
69. **Postimplantation Mammalian Embryos** Copp, A.J. & Cockroft, D.L. (Eds)
...... Spiralbound hardback 0-19-963088-7 **£15.00**
...... Paperback 0-19-963089-5 **£12.50**
68. **Receptor-Effector Coupling** Hulme, E.C. (Ed)
...... Spiralbound hardback 0-19-963094-1 **£30.00**
...... Paperback 0-19-963095-X **£19.50**
67. **Gel Electrophoresis of Proteins (2/e)** Hames, B.D. & Rickwood, D. (Eds)
...... Spiralbound hardback 0-19-963074-7 **£35.00**
...... Paperback 0-19-963075-5 **£25.00**
66. **Clinical Immunology** Gooi, H.C. & Chapel, H. (Eds)
...... Spiralbound hardback 0-19-963086-0 **£32.50**
...... Paperback 0-19-963087-9 **£22.50**
65. **Receptor Biochemistry** Hulme, E.C. (Ed)
...... Paperback 0-19-963093-3 **£25.00**
64. **Gel Electrophoresis of Nucleic Acids (2/e)** Rickwood, D. & Hames, B.D. (Eds)
...... Spiralbound hardback 0-19-963082-8 **£32.50**
...... Paperback 0-19-963083-6 **£22.50**
63. **Animal Virus Pathogenesis** Oldstone, M.B.A. (Ed)
...... Spiralbound hardback 0-19-963100-X **£15.00**
...... Paperback 0-19-963101-8 **£12.50**
62. **Flow Cytometry** Ormerod, M.G. (Ed)
...... Paperback 0-19-963053-4 **£22.50**
61. **Radioisotopes in Biology** Slater, R.J. (Ed)
...... Spiralbound hardback 0-19-963080-1 **£32.50**
...... Paperback 0-19-963081-X **£22.50**
60. **Biosensors** Cass, A.E.G. (Ed)
...... Spiralbound hardback 0-19-963046-1 **£30.00**
...... Paperback 0-19-963047-X **£19.50**
59. **Ribosomes and Protein Synthesis** Spedding, G. (Ed)
...... Spiralbound hardback 0-19-963104-2 **£15.00**
...... Paperback 0-19-963105-0 **£12.50**
58. **Liposomes** New, R.R.C. (Ed)
...... Spiralbound hardback 0-19-963076-3 **£35.00**
...... Paperback 0-19-963077-1 **£22.50**
57. **Fermentation** McNeil, B. & Harvey, L.M. (Eds)
...... Spiralbound hardback 0-19-963044-5 **£30.00**
...... Paperback 0-19-963045-3 **£19.50**
56. **Protein Purification Applications** Harris, E.L.V. & Angal, S. (Eds)
...... Spiralbound hardback 0-19-963022-4 **£30.00**
...... Paperback 0-19-963023-2 **£18.50**
55. **Nucleic Acids Sequencing** Howe, C.J. & Ward, E.S. (Eds)
...... Spiralbound hardback 0-19-963056-9 **£30.00**
...... Paperback 0-19-963057-7 **£19.50**
54. **Protein Purification Methods** Harris, E.L.V. & Angal, S. (Eds)
...... Spiralbound hardback 0-19-963002-X **£30.00**
...... Paperback 0-19-963003-8 **£22.50**
53. **Solid Phase Peptide Synthesis** Atherton, E. & Sheppard, R.C.
...... Spiralbound hardback 0-19-963066-6 **£15.00**
...... Paperback 0-19-963067-4 **£12.50**
52. **Medical Bacteriology** Hawkey, P.M. & Lewis, D.A. (Eds)
...... Paperback 0-19-963009-7 **£25.00**
51. **Proteolytic Enzymes** Beynon, R.J. & Bond, J.S. (Eds)
...... Spiralbound hardback 0-19-963058-5 **£30.00**
...... Paperback 0-19-963059-3 **£19.50**
50. **Medical Mycology** Evans, E.G.V. & Richardson, M.D. (Eds)
...... Spiralbound hardback 0-19-963010-0 **£37.50**
...... Paperback 0-19-963011-9 **£25.00**
49. **Computers in Microbiology** Bryant, T.N. & Wimpenny, J.W.T. (Eds)
...... Paperback 0-19-963015-1 **£12.50**

48.	Protein Sequencing Findlay, J.B.C. & Geisow, M.J. (Eds)		
......	Spiralbound hardback	0-19-963012-7	**£15.00**
......	Paperback	0-19-963013-5	**£12.50**
47.	Cell Growth and Division Baserga, R. (Ed)		
......	Spiralbound hardback	0-19-963026-7	**£15.00**
......	Paperback	0-19-963027-5	**£12.50**
46.	Protein Function Creighton, T.E. (Ed)		
......	Spiralbound hardback	0-19-963006-2	**£32.50**
......	Paperback	0-19-963007-0	**£22.50**
45.	Protein Structure Creighton, T.E. (Ed)		
......	Spiralbound hardback	0-19-963000-3	**£32.50**
......	Paperback	0-19-963001-1	**£22.50**
44.	Antibodies: Volume II Catty, D. (Ed)		
......	Spiralbound hardback	0-19-963018-6	**£30.00**
......	Paperback	0-19-963019-4	**£19.50**
43.	HPLC of Macromolecules Oliver, R.W.A. (Ed)		
......	Spiralbound hardback	0-19-963020-8	**£30.00**
......	Paperback	0-19-963021-6	**£19.50**
42.	Light Microscopy in Biology Lacey, A.J. (Ed)		
......	Spiralbound hardback	0-19-963036-4	**£30.00**
......	Paperback	0-19-963037-2	**£19.50**
41.	Plant Molecular Biology Shaw, C.H. (Ed)		
......	Paperback	1-85221-056-7	**£12.50**
40.	Microcomputers in Physiology Fraser, P.J. (Ed)		
......	Spiralbound hardback	1-85221-129-6	**£15.00**
......	Paperback	1-85221-130-X	**£12.50**
39.	Genome Analysis Davies, K.E. (Ed)		
......	Spiralbound hardback	1-85221-109-1	**£30.00**
......	Paperback	1-85221-110-5	**£18.50**
38.	Antibodies: Volume I Catty, D. (Ed)		
......	Paperback	0-947946-85-3	**£19.50**
37.	Yeast Campbell, I. & Duffus, J.H. (Eds)		
......	Paperback	0-947946-79-9	**£12.50**
36.	Mammalian Development Monk, M. (Ed)		
......	Hardback	1-85221-030-3	**£15.00**
......	Paperback	1-85221-029-X	**£12.50**
35.	Lymphocytes Klaus, G.G.B. (Ed)		
......	Hardback	1-85221-018-4	**£30.00**
34.	Lymphokines and Interferons Clemens, M.J., Morris, A.G. & others (Eds)		
......	Paperback	1-85221-035-4	**£12.50**
33.	Mitochondria Darley-Usmar, V.M., Rickwood, D. & others (Eds)		
......	Hardback	1-85221-034-6	**£32.50**
......	Paperback	1-85221-033-8	**£22.50**
32.	Prostaglandins and Related Substances Benedetto, C., McDonald-Gibson, R.G. & others (Eds)		
......	Hardback	1-85221-032-X	**£15.00**
......	Paperback	1-85221-031-1	**£12.50**
31.	DNA Cloning: Volume III Glover, D.M. (Ed)		
......	Hardback	1-85221-049-4	**£15.00**
......	Paperback	1-85221-048-6	**£12.50**
30.	Steroid Hormones Green, B. & Leake, R.E. (Eds)		
......	Paperback	0-947946-53-5	**£19.50**
29.	Neurochemistry Turner, A.J. & Bachelard, H.S. (Eds)		
......	Hardback	1-85221-028-1	**£15.00**
......	Paperback	1-85221-027-3	**£12.50**
28.	Biological Membranes Findlay, J.B.C. & Evans, W.H. (Eds)		
......	Hardback	0-947946-84-5	**£15.00**
......	Paperback	0-947946-83-7	**£12.50**
27.	Nucleic Acid and Protein Sequence Analysis Bishop, M.J. & Rawlings, C.J. (Eds)		
......	Hardback	1-85221-007-9	**£35.00**
......	Paperback	1-85221-006-0	**£25.00**
26.	Electron Microscopy in Molecular Biology Sommerville, J. & Scheer, U. (Eds)		
......	Hardback	0-947946-64-0	**£15.00**
......	Paperback	0-947946-54-3	**£12.50**
25.	Teratocarcinomas and Embryonic Stem Cells Robertson, E.J. (Ed)		
......	Paperback	1-85221-004-4	**£19.50**
24.	Spectrophotometry and Spectrofluorimetry Harris, D.A. & Bashford, C.L. (Eds)		
......	Hardback	0-947946-69-1	**£15.00**
......	Paperback	0-947946-46-2	**£12.50**
23.	Plasmids Hardy, K.G. (Ed)		
......	Paperback	0-947946-81-0	**£12.50**
22.	Biochemical Toxicology Snell, K. & Mullock, B. (Eds)		
......	Paperback	0-947946-52-7	**£12.50**
19.	Drosophila Roberts, D.B. (Ed)		
......	Hardback	0-947946-66-7	**£32.50**
......	Paperback	0-947946-45-4	**£22.50**

17.	Photosynthesis: Energy Transduction Hipkins, M.F. & Baker, N.R. (Eds)		
......	Hardback	0-947946-63-2	**£15.00**
......	Paperback	0-947946-51-9	**£12.50**
16.	Human Genetic Diseases Davies, K.E. (Ed)		
......	Hardback	0-947946-76-4	**£15.00**
......	Paperback	0-947946-75-6	**£12.50**
14.	Nucleic Acid Hybridisation Hames, B.D. & Higgins, S.J. (Eds)		
......	Hardback	0-947946-61-6	**£15.00**
......	Paperback	0-947946-23-3	**£12.50**
13.	Immobilised Cells and Enzymes Woodward, J. (Ed)		
......	Hardback	0-947946-60-8	**£15.00**
12.	Plant Cell Culture Dixon, R.A. (Ed)		
......	Paperback	0-947946-22-5	**£19.50**
11a.	DNA Cloning: Volume I Glover, D.M. (Ed)		
......	Paperback	0-947946-18-7	**£12.50**
11b.	DNA Cloning: Volume II Glover, D.M. (Ed)		
......	Paperback	0-947946-19-5	**£12.50**
10.	Virology Mahy, B.W.J. (Ed)		
......	Paperback	0-904147-78-9	**£19.50**
9.	Affinity Chromatography Dean, P.D.G., Johnson, W.S. & others (Eds)		
......	Paperback	0-904147-71-1	**£19.50**
7.	Microcomputers in Biology Ireland, C.R. & Long, S.P. (Eds)		
......	Paperback	0-904147-57-6	**£18.00**
6.	Oligonucleotide Synthesis Gait, M.J. (Ed)		
......	Paperback	0-904147-74-6	**£18.50**
5.	Transcription and Translation Hames, B.D. & Higgins, S.J. (Eds)		
......	Paperback	0-904147-52-5	**£12.50**
3.	Iodinated Density Gradient Media Rickwood, D. (Ed)		
......	Paperback	0-904147-51-7	**£12.50**

Sets

Essential Molecular Biology: 2 vol set Brown, T.A. (Ed)		
Spiralbound hardback 0-19-963114-X **£58.00**		
Paperback 0-19-963115-8 **£40.00**		
Antibodies: 2 vol set Catty, D. (Ed)		
Paperback 0-19-963063-1 **£33.00**		
Cellular and Molecular Neurobiology: 2 vol set Chad, J. & Wheal, H. (Eds)		
Spiralbound hardback 0-19-963255-3 **£56.00**		
Paperback 0-19-963254-5 **£38.00**		
Protein Structure and Protein Function: 2 vol set Creighton, T.E. (Ed)		
Spiralbound hardback 0-19-963064-X **£55.00**		
Paperback 0-19-963065-8 **£38.00**		
DNA Cloning: 2 vol set Glover, D.M. (Ed)		
Paperback 1-85221-069-9 **£30.00**		
Molecular Plant Pathology: 2 vol set Gurr, S.J., McPherson, M.J. & others (Eds)		
Spiralbound hardback 0-19-963354-1 **£56.00**		
Paperback 0-19-963353-3 **£37.00**		
Protein Purification Methods, and Protein Purification Applications: 2 vol set Harris, E.L.V. & Angal, S. (Eds)		
Spiralbound hardback 0-19-963048-8 **£48.00**		
Paperback 0-19-963049-6 **£32.00**		
Diagnostic Molecular Pathology: 2 vol set Herrington, C.S. & McGee, J. O'D. (Eds)		
Spiralbound hardback 0-19-963241-3 **£54.00**		
Paperback 0-19-963240-5 **£35.00**		
RNA Processing: 2 vol set Higgins, S.J. & Hames, B.D. (Eds)		
Spiralbound hardback 0-19-963473-4 **£54.00**		
Paperback 0-19-963472-6 **£35.00**		
Receptor Biochemistry; Receptor-Effector Coupling; Receptor-Ligand Interactions: 3 vol set Hulme, E.C. (Ed)		
Paperback 0-19-963097-6 **£62.50**		
Human Cytogenetics: 2 vol set (2/e) Rooney, D.E. & Czepulkowski, B.H. (Eds)		
Hardback 0-19-963314-2 **£58.50**		
Paperback 0-19-963313-4 **£40.50**		
Behavioural Neuroscience: 2 vol set Sahgal, A. (Ed)		
Spiralbound hardback 0-19-963460-2 **£58.00**		
Paperback 0-19-963459-9 **£40.00**		
Peptide Hormone Secretion/Peptide Hormone Action: 2 vol set Siddle, K. & Hutton, J.C. (Eds)		
Spiralbound hardback 0-19-963072-0 **£55.00**		
Paperback 0-19-963073-9 **£38.00**		

ORDER FORM for UK, Europe and Rest of World

(Excluding USA and Canada)

Qty	ISBN	Author	Title	Amount
			P&P	
			*VAT	
			TOTAL	

Please add postage and packing: £1.75 for UK orders under £20; £2.75 for UK orders over £20; overseas orders add 10% of total.

* EC customers please note that VAT must be added (excludes UK customers)

Name ....

Address ....

....

.... Post code ....

[ ] Please charge £ .... to my credit card

Access/VISA/Eurocard/AMEX/Diners Club (circle appropriate card)

Card No .... Expiry date ....

Signature ....

Credit card account address if different from above:

....

.... Postcode ....

[ ] I enclose a cheque for £.....................

Please return this form to: OUP Distribution Services, Saxon Way West, Corby, Northants NN18 9ES, UK

OR ORDER BY CREDIT CARD HOTLINE: Tel +44-(0)536-741519 or
Fax +44-(0)536-746337

ORDER OTHER TITLES OF INTEREST TODAY

128. **Behavioural Neuroscience: Volume I** Sahgal, A. (Ed)
...... Spiralbound hardback 0-19-963368-1 **$57.00**
...... Paperback 0-19-963367-3 **$37.00**
127. **Molecular Virology** Davison, A.J. & Elliott, R.M. (Eds)
...... Spiralbound hardback 0-19-963358-4 **$49.00**
...... Paperback 0-19-963357-6 **$32.00**
126. **Gene Targeting** Joyner, A.L. (Ed)
...... Spiralbound hardback 0-19-963407-6 **$49.00**
...... Paperback 0-19-9634036-8 **$34.00**
124. **Human Genetic Disease Analysis (2/e)** Davies, K.E. (Ed)
...... Spiralbound hardback 0-19-963309-6 **$54.00**
...... Paperback 0-19-963308-8 **$33.00**
123. **Protein Phosphorylation** Hardie, D.G. (Ed)
...... Spiralbound hardback 0-19-963306-1 **$65.00**
...... Paperback 0-19-963305-3 **$45.00**
122. **Immunocytochemistry** Beesley, J. (Ed)
...... Spiralbound hardback 0-19-963270-7 **$62.00**
...... Paperback 0-19-963269-3 **$42.00**
121. **Tumour Immunobiology** Gallagher, G., Rees, R.C. & others (Eds)
...... Spiralbound hardback 0-19-963370-3 **$72.00**
...... Paperback 0-19-963369-X **$50.00**
120. **Transcription Factors** Latchman, D.S. (Ed)
...... Spiralbound hardback 0-19-963342-8 **$48.00**
...... Paperback 0-19-963341-X **$31.00**
119. **Growth Factors** McKay, I. & Leigh, I. (Eds)
...... Spiralbound hardback 0-19-963360-6 **$48.00**
...... Paperback 0-19-963359-2 **$31.00**
118. **Histocompatibility Testing** Dyer, P. & Middleton, D. (Eds)
...... Spiralbound hardback 0-19-963364-9 **$60.00**
...... Paperback 0-19-963363-0 **$41.00**
117. **Gene Transcription** Hames, B.D. & Higgins, S.J. (Eds)
...... Spiralbound hardback 0-19-963292-8 **$72.00**
...... Paperback 0-19-963291-X **$50.00**
116. **Electrophysiology** Wallis, D.I. (Ed)
...... Spiralbound hardback 0-19-963348-7 **$56.00**
...... Paperback 0-19-963347-9 **$39.00**
115. **Biological Data Analysis** Fry, J.C. (Ed)
...... Spiralbound hardback 0-19-963340-1 **$80.00**
...... Paperback 0-19-963339-8 **$60.00**
114. **Experimental Neuroanatomy** Bolam, J.P. (Ed)
...... Spiralbound hardback 0-19-963326-6 **$59.00**
...... Paperback 0-19-963325-8 **$39.00**
113. **Preparative Centrifugation** Rickwood, D. (Ed)
...... Spiralbound hardback 0-19-963208-1 **$78.00**
...... Paperback 0-19-963211-1 **$44.00**
111. **Haemopoiesis** Testa, N.G. & Molineux, G. (Eds)
...... Spiralbound hardback 0-19-963366-5 **$59.00**
...... Paperback 0-19-963365-7 **$39.00**
110. **Pollination Ecology** Dafni, A.
...... Spiralbound hardback 0-19-963299-5 **$56.95**
...... Paperback 0-19-963298-7 **$39.95**
109. **In Situ Hybridization** Wilkinson, D.G. (Ed)
...... Spiralbound hardback 0-19-963328-2 **$58.00**
...... Paperback 0-19-963327-4 **$36.00**
108. **Protein Engineering** Rees, A.R., Sternberg, M.J.E. & others (Eds)
...... Spiralbound hardback 0-19-963139-5 **$64.00**
...... Paperback 0-19-963138-7 **$44.00**

107. **Cell-Cell Interactions** Stevenson, B.R., Gallin, W.J. & others (Eds)
...... Spiralbound hardback 0-19-963319-3 **$55.00**
...... Paperback 0-19-963318-5 **$38.00**
106. **Diagnostic Molecular Pathology: Volume I** Herrington, C.S. & McGee, J. O'D. (Eds)
...... Spiralbound hardback 0-19-963237-5 **$50.00**
...... Paperback 0-19-963236-7 **$33.00**
105. **Biomechanics-Materials** Vincent, J.F.V. (Ed)
...... Spiralbound hardback 0-19-963223-5 **$70.00**
...... Paperback 0-19-963222-7 **$50.00**
104. **Animal Cell Culture (2/e)** Freshney, R.I. (Ed)
...... Spiralbound hardback 0-19-963212-X **$55.00**
...... Paperback 0-19-963213-8 **$35.00**
103. **Molecular Plant Pathology: Volume II** Gurr, S.J., McPherson, M.J. & others (Eds)
...... Spiralbound hardback 0-19-963352-5 **$65.00**
...... Paperback 0-19-963351-7 **$45.00**
102. **Signal Transduction** Milligan, G. (Ed)
...... Spiralbound hardback 0-19-963296-0 **$60.00**
...... Paperback 0-19-963295-2 **$38.00**
101. **Protein Targeting** Magee, A.I. & Wileman, T. (Eds)
...... Spiralbound hardback 0-19-963206-5 **$75.00**
...... Paperback 0-19-963210-3 **$50.00**
100. **Diagnostic Molecular Pathology: Volume II: Cell and Tissue Genotyping** Herrington, C.S. & McGee, J.O'D. (Eds)
...... Spiralbound hardback 0-19-963239-1 **$60.00**
...... Paperback 0-19-963238-3 **$39.00**
99. **Neuronal Cell Lines** Wood, J.N. (Ed)
...... Spiralbound hardback 0-19-963346-0 **$68.00**
...... Paperback 0-19-963345-2 **$48.00**
98. **Neural Transplantation** Dunnett, S.B. & Björklund, A. (Eds)
...... Spiralbound hardback 0-19-963286-3 **$69.00**
...... Paperback 0-19-963285-5 **$42.00**
97. **Human Cytogenetics: Volume II: Malignancy and Acquired Abnormalities (2/e)** Rooney, D.E. & Czepulkowski, B.H. (Eds)
...... Spiralbound hardback 0-19-963290-1 **$75.00**
...... Paperback 0-19-963289-8 **$50.00**
96. **Human Cytogenetics: Volume I: Constitutional Analysis (2/e)** Rooney, D.E. & Czepulkowski, B.H. (Eds)
...... Spiralbound hardback 0-19-963288-X **$75.00**
...... Paperback 0-19-963287-1 **$50.00**
95. **Lipid Modification of Proteins** Hooper, N.M. & Turner, A.J. (Eds)
...... Spiralbound hardback 0-19-963274-X **$75.00**
...... Paperback 0-19-963273-1 **$50.00**
94. **Biomechanics-Structures and Systems** Biewener, A.A. (Ed)
...... Spiralbound hardback 0-19-963268-5 **$85.00**
...... Paperback 0-19-963267-7 **$50.00**
93. **Lipoprotein Analysis** Converse, C.A. & Skinner, E.R. (Eds)
...... Spiralbound hardback 0-19-963192-1 **$65.00**
...... Paperback 0-19-963231-6 **$42.00**
92. **Receptor-Ligand Interactions** Hulme, E.C. (Ed)
...... Spiralbound hardback 0-19-963090-9 **$75.00**
...... Paperback 0-19-963091-7 **$50.00**
91. **Molecular Genetic Analysis of Populations** Hoelzel, A.R. (Ed)
...... Spiralbound hardback 0-19-963278-2 **$65.00**
...... Paperback 0-19-963277-4 **$45.00**

36.	**Mammalian Development** Monk, M. (Ed)		
......	Hardback	1-85221-030-3	**$60.00**
......	Paperback	1-85221-029-X	**$45.00**
35.	**Lymphocytes** Klaus, G.G.B. (Ed)		
......	Hardback	1-85221-018-4	**$54.00**
34.	**Lymphokines and Interferons** Clemens, M.J., Morris, A.G. & others (Eds)		
......	Paperback	1-85221-035-4	**$44.00**
33.	**Mitochondria** Darley-Usmar, V.M., Rickwood, D. & others (Eds)		
......	Hardback	1-85221-034-6	**$65.00**
......	Paperback	1-85221-033-8	**$45.00**
32.	**Prostaglandins and Related Substances** Benedetto, C., McDonald-Gibson, R.G. & others (Eds)		
......	Hardback	1-85221-032-X	**$58.00**
......	Paperback	1-85221-031-1	**$38.00**
31.	**DNA Cloning: Volume III** Glover, D.M. (Ed)		
......	Hardback	1-85221-009-4	**$56.00**
......	Paperback	1-85221-048-6	**$36.00**
30.	**Steroid Hormones** Green, B. & Leake, R.E. (Eds)		
......	Paperback	0-947946-53-5	**$40.00**
29.	**Neurochemistry** Turner, A.J. & Bachelard, H.S. (Eds)		
......	Hardback	1-85221-028-1	**$56.00**
......	Paperback	1-85221-027-3	**$36.00**
28.	**Biological Membranes** Findlay, J.B.C. & Evans, W.H. (Eds)		
......	Hardback	0-947946-84-5	**$54.00**
......	Paperback	0-947946-83-7	**$36.00**
27.	**Nucleic Acid and Protein Sequence Analysis** Bishop, M.J. & Rawlings, C.J. (Eds)		
......	Hardback	1-85221-007-9	**$66.00**
......	Paperback	1-85221-006-0	**$44.00**
26.	**Electron Microscopy in Molecular Biology** Sommerville, J. & Scheer, U. (Eds)		
......	Hardback	0-947946-64-0	**$54.00**
......	Paperback	0-947946-54-3	**$40.00**
24.	**Spectrophotometry and Spectrofluorimetry** Harris, D.A. & Bashford, C.L. (Eds)		
......	Hardback	0-947946-69-1	**$56.00**
......	Paperback	0-947946-46-2	**$39.95**
23.	**Plasmids** Hardy, K.G. (Ed)		
......	Paperback	0-947946-81-0	**$36.00**
22.	**Biochemical Toxicology** Snell, K. & Mullock, B. (Eds)		
......	Paperback	0-947946-52-7	**$40.00**
19.	**Drosophila** Roberts, D.B. (Ed)		
......	Hardback	0-947946-66-7	**$67.50**
......	Paperback	0-947946-45-4	**$46.00**
17.	**Photosynthesis: Energy Transduction** Hipkins, M.F. & Baker, N.R. (Eds)		
......	Hardback	0-947946-63-2	**$54.00**
......	Paperback	0-947946-51-9	**$36.00**
16.	**Human Genetic Diseases** Davies, K.E. (Ed)		
......	Hardback	0-947946-76-4	**$60.00**
......	Paperback	0-947946-75-6	**$34.00**
14.	**Nucleic Acid Hybridisation** Hames, B.D. & Higgins, S.J. (Eds)		
......	Hardback	0-947946-61-6	**$60.00**
......	Paperback	0-947946-23-3	**$36.00**
12.	**Plant Cell Culture** Dixon, R.A. (Ed)		
......	Paperback	0-947946-22-5	**$36.00**

11a.	**DNA Cloning: Volume I** Glover, D.M. (Ed)		
......	Paperback	0-947946-18-7	**$36.00**
11b.	**DNA Cloning: Volume II** Glover, D.M. (Ed)		
......	Paperback	0-947946-19-5	**$36.00**
10.	**Virology** Mahy, B.W.J. (Ed)		
......	Paperback	0-904147-78-9	**$40.00**
9.	**Affinity Chromatography** Dean, P.D.G., Johnson, W.S. & others (Eds)		
......	Paperback	0-904147-71-1	**$36.00**
7.	**Microcomputers in Biology** Ireland, C.R. & Long, S.P. (Eds)		
......	Paperback	0-904147-57-6	**$36.00**
6.	**Oligonucleotide Synthesis** Gait, M.J. (Ed)		
......	Paperback	0-904147-74-6	**$38.00**
5.	**Transcription and Translation** Hames, B.D. & Higgins, S.J. (Eds)		
......	Paperback	0-904147-52-5	**$38.00**
3.	**Iodinated Density Gradient Media** Rickwood, D. (Ed)		
......	Paperback	0-904147-51-7	**$36.00**

Sets

	Essential Molecular Biology: 2 vol set Brown, T.A. (Ed)		
......	Spiralbound hardback	0-19-963114-X	**$118.00**
......	Paperback	0-19-963115-8	**$78.00**
	Antibodies: 2 vol set Catty, D. (Ed)		
......	Paperback	0-19-963063-1	**$70.00**
	Cellular and Molecular Neurobiology: 2 vol set Chad, J. & Wheal, H. (Eds)		
......	Spiralbound hardback	0-19-963255-3	**$133.00**
......	Paperback	0-19-963254-5	**$79.00**
	Protein Structure and Protein Function: 2 vol set Creighton, T.E. (Ed)		
......	Spiralbound hardback	0-19-963064-X	**$114.00**
......	Paperback	0-19-963065-8	**$80.00**
	DNA Cloning: 2 vol set Glover, D.M. (Ed)		
......	Paperback	1-85221-069-9	**$92.00**
	Molecular Plant Pathology: 2 vol set Gurr, S.J., McPherson, M.J. & others (Eds)		
......	Spiralbound hardback	0-19-963354-1	**$110.00**
......	Paperback	0-19-963353-3	**$75.00**
	Protein Purification Methods, and Protein Purification Applications: 2 vol set Harris, E.L.V. & Angal, S. (Eds)		
......	Spiralbound hardback	0-19-963048-8	**$98.00**
......	Paperback	0-19-963049-6	**$68.00**
	Diagnostic Molecular Pathology: 2 vol set Herrington, C.S. & McGee, J. O'D. (Eds)		
......	Spiralbound hardback	0-19-963241-3	**$105.00**
......	Paperback	0-19-963240-5	**$69.00**
	Receptor Biochemistry; Receptor-Effector Coupling; Receptor-Ligand Interactions: 3 vol set Hulme, E.C. (Ed)		
......	Paperback	0-19-963097-6	**$130.00**
	Human Cytogenetics: (2/e): 2 vol set Rooney, D.E. & Czepulkowski, B.H. (Eds)		
......	Hardback	0-19-963314-2	**$130.00**
......	Paperback	0-19-963313-4	**$90.00**
	Peptide Hormone Secretion/Peptide Hormone Action: 2 vol set Siddle, K. & Hutton, J.C. (Eds)		
......	Spiralbound hardback	0-19-963072-0	**$135.00**
......	Paperback	0-19-963073-9	**$90.00**

ORDER FORM for USA and Canada

Qty	ISBN	Author	Title	Amount
			S&H	
CA and NC residents add appropriate sales tax				
			TOTAL	

Please add shipping and handling: US $2.50 for first book, (US $1.00 each book thereafter)

Name ....

Address ....

....

.... Zip ....

[ ] Please charge $ .... to my credit card
Mastercard/VISA/American Express (circle appropriate card)

Acct. .... Expiry date ....

Signature ....

Credit card account address if different from above:

....

.... Zip ....

[ ] I enclose a cheque for US $............

Mail orders to: Order Dept. Oxford University Press, 2001 Evans Road, Cary, NC 27513